OBLIGATE ARMY-ANT-FOLLOWING BIRDS: A STUDY OF ECOLOGY, SPATIAL MOVEMENT PATTERNS, AND BEHAVIOR IN AMAZONIAN PERU

ORNITHOLOGICAL MONOGRAPHS

Editor: John Faaborg

224 Tucker Hall
Division of Biological Sciences
University of Missouri
Columbia, Missouri 65211

Managing Editor: Bradley R. Plummer
Proof Editors: Mark C. Penrose, Richard D. Earles

AOU Publications Office
622 Science Engineering
Department of Biological Sciences
University of Arkansas
Fayetteville, Arkansas 72701

The *Ornithological Monographs* series, published by the American Ornithologists' Union, has been established for major papers too long for inclusion in the Union's journal, *The Auk*.

Copies of *Ornithological Monographs* are available from Buteo Books, 3130 Laurel Road, Shipman, VA 22971. Price of *Ornithological Monographs* no. 55: $10.00 ($9.00 for AOU members). Add $4.00 for handling and shipping charges in U.S., and $5.00 for shipping to other countries. Make checks payable to Buteo Books.

Author of this issue, Susan K. Willson.

Library of Congress Control Number 2004105806

Printed by Cadmus Communications, Ephrata, PA 17522

Issued 5 May 2004

Ornithological Monographs, No. 55 x + 67 pp.

ISBN: 0-943610-60-5

Cover: Assemblage of ant-following birds: (1) *Dendrocincla merula*, (2) *Rhegmatorhina melanosticta*, (3) *Phlegopsis nigromaculata*, (4) *Myrmeciza fortis*, and (5) *Gymnopithys salvini*. (Ink sketch by Kirsten Carlson.)

OBLIGATE ARMY-ANT-FOLLOWING BIRDS: A STUDY OF ECOLOGY, SPATIAL MOVEMENT PATTERNS, AND BEHAVIOR IN AMAZONIAN PERU

By

Susan K. Willson

Division of Biological Sciences, University of Missouri, 105 Tucker Hall, Columbia, Missouri 65211, USA

ORNITHOLOGICAL MONOGRAPHS NO. 55

PUBLISHED BY
THE AMERICAN ORNITHOLOGISTS' UNION
WASHINGTON, D.C.
2004

TABLE OF CONTENTS

LIST OF TABLES

LIST OF FIGURES

APPENDIX

From the Editor

This is the second *Ornithological Monograph* distributed to all members of the American Ornithologists' Union. Whereas *Ornithological Monograph No. 54* narrowly focused on the demography of a single species, with contributions from multiple co-authors, this monograph by Susan K. Willson attempts to describe the ecological interactions involved in the coexistence of five species that spend nearly all their lives foraging at the front of antswarms. To appreciate the many mechanisms involved in how these birds coexist, we also need to know a fair amount about the ants that provide the foraging location used by the birds.

This monograph includes much of the research done by its author for her doctoral dissertation. Without something like this monograph series, the dissertation would have been separated into five or six units that likely would have appeared in five or six different journals. The ant ecology that is so important to her story would likely have been in a journal ornithologists rarely read. Anyone who wanted to see the whole story would have had to round up those articles from an array of locations and put the package together. Instead, with this monograph, we can read the whole story about how the various ecological factors interact to suggest how five obligate ant-followers can coexist in a single Peruvian rainforest.

We hope this monograph encourages other graduate students to consider telling the story of their own research in a single location when there is a longer, more complex story than can be told in piecemeal fashion. We also recognize the downside of publishing a single monograph rather than five or six separate publications; we hope that department chairs, deans, and provosts will recognize that a monograph ought to count for more than one publication when considerations of hiring and tenure are involved!

Reviewing and editing a dissertation that becomes a monograph involves the help of many people. Susan K. Willson's doctoral committee included John Terborgh, Bette Loiselle, Frank Thompson III, Rex Cocroft, and myself; all of these individuals made numerous comments on development of the research and early drafts of its results. Edwin O. Willis and Phillip Stouffer were kind enough to spend a large amount of their time making additional comments on the monographic form of Sue's dissertation. Kimberly Smith, Bradley Plummer, Mark Penrose, and Richard Earles of the AOU Publications Office were exceedingly helpful as we pushed deadlines. Dr. John David, Chair of the Division of Biological Sciences of the University of Missouri-Columbia, provided funds to support the color plate. We thank all of these individuals for helping to make *Ornithological Monograph No. 55* a compelling examination of some of the most interesting birds of the New World tropics.

John Faaborg

Color plate: From left to right, beginning upper left: *Rhegmatorhina melanosticta* (adult; photo copyright 2004, C. E. T. Paine); *Gymnopithys salvini* (female); *Myrmeciza fortis* (female); *Rhegmatorhina melanosticta* (female); *Dendrocincla merula* (adult); *Eciton burchelli* army ants; *Myrmeciza fortis* (male); *Phlegopsis nigromaculata* (adult); *Gymnopithys salvini* (male).

Ornithological Monographs
Volume (2004), No. 55, 1–67

OBLIGATE ARMY-ANT-FOLLOWING BIRDS: A STUDY OF ECOLOGY, SPATIAL MOVEMENT PATTERNS, AND BEHAVIOR IN AMAZONIAN PERU

Susan K. Willson[1]

Division of Biological Sciences, University of Missouri, 105 Tucker Hall, Columbia, Missouri 65211, USA

Abstract.—Five species of obligate ant-following birds—*Phlegopsis nigromaculata, Myrmeciza fortis, Rhegmatorhina melanosticta, Gymnopithys salvini* (Thamnophilidae), and *Dendrocincla merula* (Dendrocolaptidae)—and two species of army ants (*Eciton burchelli* and *Labidus praedator*) were studied in Amazonian Peru over five years. Here, I explore aspects of species coexistence in these five ecologically similar birds through analyses of their population ecology, resource use, behavior, and spatial movement patterns.

Mean home-range size of each antbird species was reduced through reliance on the unpredictable but abundant foraging opportunities at *L. praedator* swarms. This little-known ant species played a pivotal role in expanding the foraging resource available to the obligate ant-following birds, which allowed an increase in the birds' population densities well above what would be supported solely by the better-known *E. burchelli* army ants.

Two of the five bird species (*D. merula* and *M. fortis*) displayed resource selectivity among antswarms by foraging significantly more with one of the two ant species. The woodcreeper *D. merula* further segregated from the four antbirds in its utilization of white-lipped peccary (*Tayassu peccari*) herds as a foraging resource; the peccaries act as "beaters" of arthropod prey in a manner similar to that of the army ants. The three antbirds that did not prefer one army ant species over the other (*P. nigromaculata, R. melanosticta,* and *G. salvini*) segregated by body mass, which may allow differential use of space along the width of an antswarm front. That size difference would permit a smaller, more subordinate species to "fit" along the front of a swarm that was already "full" to a different bird species.

Population dynamics of the birds were not stable over five years of data collection, and total population of obligate ant-followers declined by almost half over the course of the study. It is suggested that periodic population fluctuations are a normal occurrence in guilds of obligate ant-followers and may be exacerbated by the lack of territoriality exhibited by most of these species. Lower population density correlated with decreased interference competition among individuals. Population fluctuations may increase the ability of the subordinate species *R. melanosticta* to coexist with the larger, dominant *P. nigromaculata* in floodplain forest.

Nest-site selection may contribute to niche breadth among the obligate ant-followers. I provide descriptions of the nests, eggs, and nestlings of *P. nigromaculata, R. melanosticta,* and *G. salvini,* which were undescribed at the start of the present study. *Received 22 November 2003, accepted 3 February 2004.*

Resumen.—Se estudiaron cinco especies de aves que siguen hormigas arrieras—*Phlegopsis nigromaculata, Myrmeciza fortis, Rhegmatorhina melanosticta* y *Gymnopithys salvini* (Thamnophilidae) y *Dendrocincla merula* (Dendrocolaptidae)—y dos especies de estas hormigas (*Eciton burchelli* y *Labidus praedator*) en la Amazonía peruana durante cinco años. En este trabajo se exploran aspectos de la coexistencia de estas cinco especies ecológicamente similares por medio de análisis de su ecología poblacional, uso de recursos, comportamiento y patrones espaciales de movimiento.

El tamaño promedio del rango de hogar de cada especie de Thamnophilidae fue reducido por su dependencia en las impredecibles pero abundantes oportunidades de forrajear que brindan los ejércitos de *L. praedator*. Esta especie de hormiga jugó un papel determinante para expandir los recursos disponibles para las aves, lo que permitió un incremento en la densidad poblacional de éstas más allá de los niveles que podrían mantenerse únicamente con las hormigas de la especie *E. burchelli*.

Dos de las cinco especies de aves (*D. merula* y *M. fortis*) exhibieron selectividad de recursos

[1]E-mail: susan@phlegopsis.com

entre legiones de hormigas, forrajeando signficativamente más junto a una de las dos especies de hormigas. Además, el trepatronco *D. merula* se segregó ecológicamente de las otras cuatro especies de aves en términos de su utilización de manadas de huanganas (*Tayassu peccari*) como recurso de forrajeo; estos animales espantan artrópodos que pueden ser depredados por las aves en una forma similar a como lo hacen las hormigas arrieras. Las tres especies que no prefirieron una especie de hormiga sobre la otra (*P. nigromaculata, R. melanosticta* y *G. salvini*) se segregaron en términos de su peso corporal, lo que podría permitir un uso diferencial del espacio en el frente de avance de los ejércitos de hormigas. Las diferencias de tamaño podrían permitir que las especies más pequeñas y subordinadas puedan acomodarse en el frente de un ejército que ya estaría lleno para individuos de otra especie.

Las dinámicas poblacionales de las aves no fueron estables a través de los cinco años de toma de datos, y el total de la población de aves seguidoras de hormigas declinó a casi la mitad durante el curso del estudio. Se sugiere que las fluctuaciones poblacionales periódicas son eventos normales en los gremios de aves que siguen hormigas obligatoriamente, y que éstas podrían ser acentuadas por la falta de territorialidad exhibida por la mayoría de las especies. Los tamaños poblacionales menores se correlacionaron con una competencia por interferencia reducida entre individuos. Las fluctuaciones poblacionales podrían incrementar la habilidad de la especie subordinada *R. melanosticta* para coexistir en el bosque del plano de inundación con *P. nigromaculata*, que es más grande y dominante.

La selección de sitios de nidificación podría contribuir a la amplitud del nicho entre las aves seguidoras de hormigas. Aquí proveo descripciones de los nidos, huevos y pichones de *P. nigromaculata, R. melanosticta* y *G. salvini*, los cuales no habían sido descritos al comienzo de este estudio.

1. INTRODUCTION

FOR CENTURIES, SCIENTISTS have been awed by the richness of tropical diversity. Early naturalists introduced the faunal and floral exuberance of the Amazon region to the world through their species collections and natural-history writings (von Humboldt and Bonpland 1814–1829, Bates 1862, Wallace 1889, Spruce 1908). Since the time of Wallace, scientists have remarked on a latitudinal species-diversity gradient. High species diversity in the tropics results from a combination of historical, climatic, and ecological factors and area (Dobzhansky 1950, Fischer 1960, Connell and Orias 1964, Fedorov 1966, Pianka 1966, Janzen 1967, MacArthur 1969, Connell 1978, Huston 1979, Ricklefs and Schluter 1993, Rosenzweig 1995, Gaston 1996, Leigh 1999). Within tropical guilds, higher species-packing and finer niche-partitioning along axes related to body size, foraging behavior, and vertical habitat-stratification contribute to coexistence of ecologically similar species (Hutchinson 1959, MacArthur 1969, Diamond 1975, Terborgh 1980, Sherry 1984, Fitzpatrick 1985, Grant 1985, Marra and Remsen 1997, Rosenberg 1997). Levey and Martínez del Rio (2001), for example, included physiological ability to digest fruit among their axes for avian frugivore diversification; by examining why more birds do not eat fruit, they provided new insight into patterns of frugivore coexistence.

Avian species diversity reaches its peak in the Neotropics, where some 3,300 species are found (Karr 1989). The richest region of the Neotropics may be in the lowland rainforests of southwestern Amazonia. Researchers in Manu National Park, Peru, have recorded >540 bird species within ~3 km^2 of the station (Terborgh et al. 1984). At that site, 245 species are resident territory-holders on a 97-ha floodplain forest plot. By superimposing territory maps or areas of occupancy of individual species, Terborgh et al. (1990) quantified single-point (alpha) diversity at >160 species through territory overlap.

Diversification and specialization in tropical birds is aided by resources that are available year-round in the tropics (Orians 1969). One such resource, unique to the tropics, are the army ants (Willis and Oniki 1978). Army ants forage both below ground and over leaf litter for a wide range of invertebrate prey (Rettenmeyer 1963, Schneirla 1971, Hölldobler and Wilson 1990). Most of the 144 species of army ants in the subfamily Ecitoninae forage in long columns (Hölldobler and Wilson 1990). However, two species, *Eciton burchelli* and *Labidus praedator*, forage in swarms that blanket the ground, advancing in fan-shaped raids or patches. Moving over the ground surface, they flush leaf-litter arthropods. If the ants do not catch them immediately, the insects flee by escaping ahead of the swarm front, providing an easily accessible food resource for birds. Many species of birds opportunistically use army ants as a foraging resource (Willis and Oniki 1978); >49 species have been observed foraging at swarms in Amazonian Peru (S. K. Willson unpubl. data). Some species, however, appear to feed only or mainly at antswarms. Although army ants offer a predictable resource in the long term, swarms are spatially patchy and unpredictable in the short term (Schneirla 1971). Specialists (or "obligates") on army ants must overcome the challenges of finding and tracking swarms from day to day and of foraging effectively amid numerous and potentially larger competitors.

The obligate ant-following guild is ecologically interesting, because it appears that several coexisting obligate species are sharing the same fundamental ecological niche (Grinnell 1922, Gause 1934, MacArthur 1958, Hutchinson 1959). At some sites, up to three obligate species may be of equal body mass (Terborgh et al. 1990). Additionally, diet composition and prey size have been shown to completely overlap among obligate ant-following guild members, countering predictions of species coexistence based on body size and interspecific competition (Chesser 1995). Willis and Oniki (1978) described a series of concentric feeding zones in a simple ant-following community in Panama, in which the dominant (i.e. larger) obligate species monopolized areas closer to flushed prey. Willis's numerous studies indicate that ant-following species may segregate by perch type at swarm fronts, using vertical perches of various sizes, the ground, or horizontal perches. Additionally, Willis and Oniki (1978) suggest that obligate ant-following birds segregate by dominance based on size—with larger, more dominant species taking the lowest and closest position to the oncoming antswarm so as to capture the largest amount of fleeing arthropod prey (see summary in Zimmer and Isler 2003). Explanations of coexistence are less clear for Amazonian guilds, in which members frequently overlap in body mass, foraging behavior, or feeding preferences. Building on explanations for species coexistence in obligate ant-followers first offered by Willis and Oniki (1978), I examined differential use of available foraging resources (i.e. the army ants) at the community level, as well as temporal variation in population-level processes. I considered the behavioral and small-scale ecological patterns of individual birds and army ant colonies, and the broader patterns of species' annual survival, population dynamics, and recruitment.

2. BACKGROUND: SOME NATURAL HISTORY OF THE BIRDS AND ARMY ANTS

The Birds

Obligate ant-following birds demonstrate the evolution of a highly specialized lifestyle, which is intrinsically tied to the ecology of the army ants on which they depend for food. Obligate ant-followers seem behaviorally incapable of catching prey that is not being flushed by another organism (see Skutch 1996 for an account of the author acting as a "beater" of prey for an obligate ant-following bird). There are no published accounts of an obligate ant-follower hunting and catching prey without the service of army ants or other beaters of prey. Hence, obligate ant-followers might more correctly be termed "obligate followers of beaters." Generally, army ants provide that service. Willis and Oniki (1978) discussed "professional" and "facultative" ant-followers, but allowed that professionals may forage on their own, away from army ants, for up to 50% of the time. Here, I use the term "obligate" for all five study species (four antbirds and one woodcreeper). During five years of careful observation and intensive use of radiotelemetry, no individuals were ever known to forage away from beaters of prey. The woodcreeper was found, during the study, to obligately depend on two types of beaters to flush prey—army ants and herds of white-lipped peccaries (*Tayassu pecari*). For convenience, all five species are referred to here as "obligate ant-followers," even though the woodcreeper is technically an "obligate follower of beaters of prey" at the study site.

Few researchers have studied the natural history of obligate ant-following birds. The most significant and well-known publications on the subject come from the prolific E. O. Willis, who began his studies in Central America and subsequently traveled the Amazon, publishing an extensive literature on the natural history of members of the obligate ant-following guild (see Willis and Oniki 1978, 1992). Much of his work describes the posturing and perch use, vocalizations, intra- and interspecific interactions, and minutiae of daily life at an antswarm front. Most other published information is anecdotal or scattered, and obligate ant-followers have attained—because of their specificity of habitat and diet, the difficulty of observing them away from antswarms, and their rarity as a group—a "legendary" status (especially among birders) that is perpetuated in popular writing (Forsyth and Miyata 1984, Kricher 1997).

Here, I introduce the five study species and briefly review their social systems, which range from solitary individuals to family groups. All information not specifically referenced was first described in the present study.

The dominant obligate ant-follower at Cocha Cashu, *P. nigromaculata* (Black-spotted Bare-eye), lives in family groups of two to possibly four adults, plus juveniles in the postbreeding season, and averages 46 g at Cocha Cashu. Cadena et al. (2000) first described the nest. The only published behavioral information comes from Willis (1979), who studied the species in Brazil. Composition of the species' family groups is not well understood. At Cocha Cashu, multiple family-groups gather at antswarms, and individuals from different families may perch and "loaf" together between foraging bouts. Additionally, I found that a family may simultaneously have individuals foraging at different swarms hundreds of meters apart, as in *Phaenostictus mcleannani* (Ocellated Antbird) in Panama (Willis 1973). Willis (1979) speculated that "extra males" of *Pha. mcleannani*, probably sons of previous clutches, stay with the parents for several years before dispersing. At Cocha Cashu, I confirmed that *P. nigromaculata* is a cooperative breeder (see Chapter 5).

Although on average it is slightly larger (at 46.5 g) than *P. nigromaculata*, *M. fortis* (Sooty Antbird) is second in the strict linear-dominance hierarchy that exists among the obligate antbird species. No accounts of its natural history have been published, though Willis (1985) describes frequent ant-following by *M. fortis* and related species. Its nest was first described in 1997 (Wilkinson and Smith 1997). *Myrmeciza fortis* individuals form monogamous pairs that are highly aggressive both inter- and intraspecifically. The lack of difference in mass between *M. fortis* and the dominant *P. nigromaculata* (Table 2.1) suggests that group size may account for the subordinate status of *M. fortis*. Given that *P. nigromaculata* individuals generally forage in groups of three to five, it is possible that their sheer number, through collective resource-defense, allows for species

TABLE 2.1. Morphological measurements of five obligate ant-following bird species at Cocha Cashu Biological Station, Peru. Sample size is in parentheses. All data are weighted to correct for multiple captures per individual. All measurements are of adults only. Abbreviations: Phleg = *Phlegopsis nigromaculata*; M.fort = *Myrmeciza fortis*; D.mer = *Dendrocincla merula*; Rheg = *Rhegmatorhina melanosticta*; G.sal = *Gymnopithys salvini*.

Species	Average weight (g)	SD weight (g)	Average wing[a] (mm)	SD wing (mm)	Average tarsus (mm)	SD tarsus (mm)	Average culmen[b] (mm)	SD culmen (mm)	Average bill width[c] (mm)	SD bill width (mm)	Average bill depth[d] (mm)	SD bill depth (mm)
Phleg	46.0 (64)	3.9	86.5 (64)	3.3	31.6 (64)	2.1	12.1 (64)	1.2	4.6 (64)	0.4	5.6 (64)	0.3
M.fort	46.5 (18)	2.8	81.7 (18)	2.6	33.0 (18)	1.9	13.5 (18)	0.7	5.4 (18)	0.3	6.4 (18)	0.4
D.mer	47.5 (67)	3.4	104.4 (67)	3.0	26.4 (67)	1.9	19.1 (67)	1.5	5.5 (67)	0.4	6.5 (67)	0.4
Rheg	31.4 (41)	8.5	75.5 (41)	2.0	28.5 (41)	2.6	11.4 (41)	0.7	4.6 (41)	0.4	5.1 (41)	0.3
G.sal	25.9 (70)	2.3	71.0 (71)	2.4	27.1 (69)	2.5	11.4 (70)	0.6	4.4 (70)	0.3	4.6 (70)	0.2

[a] Wing chord measure is unflattened.
[b] Culmen length: measured from anterior edge of nares to bill tip.
[c] Width of bill: measured at anterior edge of nares.
[d] Depth of bill: measured at anterior edge of nares.

dominance over *M. fortis* pairs. Terborgh (1983) described a similar example of cooperatively exerted dominance between two *Cebus* monkey species that differ in mass. *Myrmeciza fortis* is notable for its extreme intolerance of conspecifics. Willis (1985) noted different pairs at ends of the same antswarm, but observed fights between pairs. Unlike other obligate ant-followers at Cocha Cashu, *M. fortis* pairs never shared a swarm with conspecific pairs during the present study and held all-purpose "type A" defended territories (as described in Karr 1971). All occurrences of more than two adults at a swarm during the present study were of parents with a grown offspring.

The status of *M. fortis* as an obligate ant-follower was in doubt when the present research began, because no study of its behavior had been published previously and because of its phylogenetic placement apart from the other obligate antbirds. The genus *Myrmeciza* has high morphological variability and encompasses generalist foragers, facultative ant-followers, and a few species believed to exclusively follow army ants (see Zimmer and Isler 2003). Through radiotelemetry, the present study demonstrated that *M. fortis* is an obligate ant-follower that feeds exclusively at antswarms. *Myrmeciza fortis* also engages in the specialized behavior of bivouac-checking (see Swartz 2001), though Willis (1982a and E. O. Willis pers. comm.) showed that facultative species may perform bivouac-checks as well.

The medium-sized *R. melanosticta* (Hairy-crested Antbird) is the third obligate antbird species in the dominance hierarchy at Cocha Cashu, subordinate to both *P. nigromaculata* and *M. fortis*. *Rhegmatorhina melanosticta* averaged 31.4 g and generally maintained socially monogamous pair bonds. In the only publication dealing with this species, Willis (1969) detailed behavior of the genus (five species), but he did not follow marked individuals and did not perform long-term observations of *R. melanosticta*. My work suggests that the frequency of exclusive pair bonds may be inversely related to population density. A dramatic decrease in adult population density between 1998 and 2000 corresponded with changes in observed intraspecific interactions. In the 1998 and 1999 field seasons, individuals generally formed loose associations and seemed to form exclusive pair bonds only during nesting attempts. In 2000, all adults were in tight pair bonds and were seen apart only while incubating. The lack of available mates was demonstrated when an adult male began courting a fledgling female less than one month out of the nest; the male was accepted by her family and began foraging with them daily. The male took over provisioning of the fledgling, which had still been receiving food from her parents. Alternatively, the observed population decline may have eliminated "floaters" (unpaired individuals) and decreased competition for mates, allowing all individuals to form pair bonds. When population density was high, 11 adult individuals were present at a large antswarm.

The smallest and most subordinate obligate ant-follower in the present study is *G. salvini* (White-throated Antbird). This species averaged 25.9 g and maintained socially monogamous pair bonds. Willis (1968) observed unbanded individuals of the species for less than a week; his study is the only published behavioral information for the species. As many as five pairs of *G. salvini* aggregated at large antswarms, and intraspecific aggression was common. Willis (1967) describes reversals of dominance for *G. leucaspis* in Panama, when a pair crosses into the center area of another pair. Although individual dominance interactions were noted in the present study, I did not observe clear reversals of dominance across territorial boundaries in *G. salvini*.

Finally, *D. merula* (White-chinned Woodcreeper), averaging 47.5 g, was subordinate to the two largest antbirds but received and initiated agonistic encounters with *R. melanosticta* (average 31.4 g), leaving the two species similarly positioned in the dominance hierarchy. *Dendrocincla merula* was mainly intraspecifically aggressive, and individuals sometimes lost large portions of foraging time to conspecific aggression. Those interactions generally did not preclude individuals from gathering at the same swarm; sometimes more than five adults were present at once. My observations from Cocha Cashu are counter to those of Willis (1979) in Brazil, who reported that individuals "dispute with each other little." However, *D. merula* is rare in all Brazilian areas Willis has checked (E. O. Willis pers. comm.).The undescribed nest is presumably tended solely by the female, as in other *Dendrocincla* species (Willis 1972). Individuals did not form long-term pair bonds, and females cared for fledged young alone. Males seemed to contribute sperm and little else, which suggests female choice for male quality. Overlapping home ranges and regular meetings with males at antswarms likely allow females to evaluate a number of potential mates.

Although not part of the present study, four facultative ant-followers at Cocha Cashu—*Hylophylax poecilonota, Dendrocolaptes picumnus, Dendrocincla fuliginosa,* and *Neomorphus geoffroyi*—deserve mention here (based on previous published accounts of their use of antswarms).

Hylophylax poecilonota (Scale-backed Antbird) is a small thamnophilid species (see Willis 1982c), but resides mainly on *terra firme* forest at Cocha Cashu. It was very rarely seen in the study area (three sightings in four years).

Dendrocolaptes picumnus (Black-banded Wood-creeper) attended antswarms (see Willis 1984), but its status as an obligate at Cocha Cashu was rejected by Pierpont (1986); she observed *Dendroc. picumnus* foraging both at antswarms and with understory mixed-species flocks and characterized its affinity for army ants as "medium."

Another woodcreeper species, *D. fuliginosa* (Plain-brown Woodcreeper), was characterized by Willis (1966a) as a "professional ant follower in all areas I have studied it," though he mentioned that it forages away from swarms on occasion. At Cocha Cashu, *D. fuliginosa* is a facultative ant-follower and spends most of its time with understory mixed-species flocks. From observation and radiotelemetry, Pierpont (1986) characterized the species' affinity for army ants as "low."

Neomorphus geoffroyi (Rufous-vented Ground-Cuckoo) is a species whose rarity and shyness precluded much observation of it. Almost nothing is known about its ecology or behavior (Willis 1982b, Ridgely and Greenfield 2001). It was sometimes sighted at antswarms, but was also seen foraging under squirrel monkeys (*Saimiri sciureus*), which act as beaters of prey as they drop arthropods to the ground (M. B. Swartz pers. comm.). The name given to *N. geoffroyi* by the local Matsigenka people, "huangana piscco," translates to "white-lipped-peccary follower." Foraging across the forest floor for nuts, roots, and animal matter, white-lipped peccary herds can flush arthropods; herds were regularly seen on the study site (Silman et al. 2003).

Myrmoborus myotherinus (Black-faced Antbird) was the most common facultative ant-follower at Cocha Cashu antswarms, and individuals and pairs frequented swarms of both ant species. Specifically, I recorded *Myrmo. myotherinus* at 94 of 345 observed swarms (27%) in 2000–2001. All birds in the present study were dominant to *Myrmo. myotherinus*.

Family groups of *Psophia leucoptera* (White-winged Trumpeter; Psophiidae) occasionally foraged at antswarm fronts. Individual trumpeters sometimes ran through the front to capture large fleeing arthropods (i.e. centipedes, scorpions, and tarantulas) and momentarily disrupted foraging by obligate ant-followers. Significant disruption of the front occurred only when two family groups converged at a swarm. Family groups chased each other and at times disturbed the forward movement of the advancing army ants. However, front cohesiveness was not affected for more than a few moments, and obligate ant-following birds did not leave swarms attended by trumpeters.

The Army Ants

The ecology of the Neotropical army ant *E. burchelli* (Formicidae: subfamily Ecitoninae, tribe Ecitonini) has been studied by only a handful of investigators. Much of the published information on army ant natural history comes from T. C. Schneirla, who dedicated his life's work from the 1930s to 1960s to their study (compiled in Schneirla 1971). Beginning in the 1960s, E. O. Willis amassed data during his study of the obligate antbird *G. leucaspis* (Bicolored Antbird) in Panama. Willis's data on *E. burchelli* were analyzed by Nigel Franks, who augmented them with his own while studying army ant population ecology, beginning with his doctoral dissertation (Franks 1980). Franks has advanced the understanding of army ant spatial dynamics through use of simulation and mathematical models. Schneirla, Willis, and Franks did the majority or all of their work on Barro Colorado Island in Panama. Unfortunately, lack of studies elsewhere has led to often-repeated assumptions about army ant population densities in both scientific journals and the popular press. My data from southeastern Peru demonstrate that population densities are >1.5× higher there than on Barro Colorado Island, where densities have remained stable at 3.2 colonies per km^2 for >50 years (Willis 1967, Schneirla 1971, Franks 1982a).

Eciton burchelli ants live in colonies of 500,000 to 2,000,000 individuals, and their life cycle has two distinct phases (Schneirla 1971). During the nomadic phase, which lasts ~14 days, the entire colony moves its temporary nest, or bivouac, almost every night. During the day, adult workers forage out from the bivouac in a swarm raid to capture arthropods and social-insect larvae, which they carry back to the bivouac—where the army ants' own brood of ~300,000 larvae is waiting to be fed (Fig. 2.1).

During the statary phase, the colony remains in a protected nest-site for ~21 days. The single queen lays her eggs over a period of 3–4 days, and all larvae from the previous nomadic phase pupate. Relieved of the need to feed larvae, adult workers forage sporadically for prey. At the end of the statary phase, eggs hatch into larvae, and pupae emerge from their cocoons and join the ranks of adult workers. The army ant cycle is therefore a 35-day period. An individual egg must go through two cycles (70 days) before the pupa emerges as a new, or callow, adult worker. Callow workers are lighter in color than their older adult counterparts, and thus can be visually distinguished from the latter until their exoskeletons darken, after approximately 2–3 days.

During the nomadic phase, swarm raids begin around dawn. Ant activity in the bivouac quickly changes from little or no visible activity on the part of the workers to a sudden exodus from the bivouac, which can be likened to water pouring from a waterfall. The ground immediately surrounding the bivouac becomes dark with ants; within 30 min, the ants choose a primary foraging direction and form a column pushing outward in that direction. The foraging body can be thought of in metaphor as a tree. The base of the tree is the bivouac, and the trunk is the ever-lengthening foraging column (generally only a few centimeters wide). Toward the distal end of the column, workers begin a process of dendritic branching, forward and outward from the column. The branches become smaller along the column until all branches merge at the swarm front (the top of the "tree"), where ants (the "leaves") darken the ground. The swarm front varies in size depending on time of day, colony size, phase, and other factors, but can reach >25 m across at Cocha Cashu. The swarm front is where the ants are actively searching for and capturing prey. As they move through the leaf litter, arthropods and vertebrates in their path hop, fly, or run ahead of the advancing swarm. In dry leaf litter, the noise of fleeing prey is quite audible.

FIG. 2.1. *Eciton burchelli* army ants build nests, or bivouacs, entirely from their own bodies. The image above shows a nomadic bivouac ~1.5 m long, under a fallen log. The image on the left is a close-up of that bivouac, with worker ants moving out from the bivouac to forage.

In addition to preying on ground fauna, army ants climb trees in their path to scout for nests of social insects. Rettenmeyer (1963) has shown that social-insect larvae, particularly those of other ant species, are the preferred prey of *E. burchelli* and generally account for the majority of their prey intake. When swarm raids find a large social-insect nest in a tree, the entire front may stop its ground raid and concentrate its efforts on retrieving larvae from that nest. Adults of the raided nests soon give up any attempt to rebuff the army ants, and often attempt to save themselves by "raining" out of an *E. burchelli*-filled tree onto the forest floor below.

Traffic along the *E. burchelli* foraging column is continuously bidirectional and connected to the bivouac. Because army ants forage forward from the bivouac at ~14 m h^{-1}, the foraging front can be >125 m from the bivouac by late afternoon. During the nomadic phase, a colony will generally move its entire bivouac—beginning at the end of the day and working through the night—to a new location. The ants place the new bivouac somewhere along that day's exact foraging path by following the pheromone trail. At times, a colony may forage in more than one direction from the bivouac, either simultaneously or over the course of a day. That is a more common phenomenon during the statary phase, but can also occur during the nomadic phase when there is an obstruction in the primary direction (e.g. a body of water, other *E. burchelli* ant-colony pheromones, a rain shower).

A second army ant species, *L. praedator*, is also an important resource for the obligate ant-followers at Cocha Cashu. Not as well studied as *E. burchelli* because of their subterrenean habits, *L. praedator* are often called "rain ants," from their propensity to swarm on the forest floor after a downpour. They are driven to the surface by inundation of the subterranean cavities they generally favor for raiding; thus, the species often exhibits a seasonal pattern of aboveground swarming. E. O. Willis (pers. comm.) has speculated that *L. praedator* may forage more nocturnally during dry weather. *Labidus praedator* does not display the cyclic lifestyle of *E. burchelli*, and can remain at the same nest-site for up to five months (Schneirla 1971). Few studies of its ecology have been undertaken, but Schneirla (1971) and Rettenmeyer (1963) provided much basic natural history and biology. Individual *L. praedator* colonies are a spatially and temporally unreliable resource for obligate ant-following birds, because the location and activity of aboveground swarms are unpredictable. When they do raid above ground, swarms are small (average width 2.5 m), move in an "S" shape with frequent changes of direction, and can disappear underground at any moment. However, *L. praedator* swarms are a valuable resource for obligate ant-following birds that find them opportunistically, because they are much more common at Cocha Cashu than *E. burchelli* swarms.

The Study Site

Field data were collected at Cocha Cashu Biological Station, Manu National Park, Peru (11°54'S and 71°22'W; elevation ~400 m). Mean annual temperature is 23.5°C, and annual rainfall averages 2,000 mm, putting the site near the climatic boundary between tropical and subtropical moist forest in the Holdridge system (Holdridge 1967, Terborgh 1990). The dry season lasts from April to mid-September, and the rainy season from September or October through March. The station and surrounding area lie in the floodplain of the Manu river, a meandering whitewater Amazonian tributary of the Rio Madre de Diós. The forest has a canopy height of ~30 m, with scattered emergent trees reaching 50 m in height. Large tree-falls are common, because the soils are regularly inundated. The forest is in a constant state of succession, new tree-falls creating light-gaps and old tree-falls filling with fast-growing saplings and lianas. Interspersed within the forest are lower areas that become inundated during the rainy season, creating a mosaic of dry and wet patches (0.1 to 1.5 m deep) that may persist until the following dry season. Larger depressed areas within the study plot are fig swamps, formed when old oxbow lakes dried up and filled with sediment. Those areas are characterized by scattered, sprawling fig trees (*Ficus trigona*) and other canopy cover, with few understory trees but a dense undergrowth of tall herbaceous plants (*Heliconia* spp. and Marantaceae). A detailed description of the ecological, geological, and climatic characteristics of the site can be found in Terborgh (1983) and Gentry (1990).

3. RESOURCE USE AND SPECIES COEXISTENCE IN FIVE OBLIGATE ANT-FOLLOWING BIRD SPECIES

Species that occupy similar ecological niches within a habitat will compete for resources that are in limited supply (MacArthur 1958, Connell 1961). Interspecific competition may negatively affect fecundity, survival, or growth of one or more competing species through density-dependent effects (Feinsinger 1976; Brown 1989a, b). Theoretical, experimental, and empirical studies have shown that for ecologically similar species to coexist in a habitat with limited resources, niche differentiation must occur (MacArthur and Levins 1964; May 1973; Schoener 1974; Brown 1975; Brown 1989a, b). Niche differentiation may involve dissimilarities in body size, habitat use, food choice, or temporal activity patterns (Lack 1944, Hutchinson 1957, Schoener 1974, Terborgh 1983, Durant 1998, Himes 2003). In natural communities, a heterogeneous environment is coupled with evolutionary trade-offs among coexisting species in their abilities to use various parts of the environment. Each species may have certain behavioral adaptations that provide advantages over its competitors under specific combinations of environmental conditions, allowing competing species to coexist in the same environment (MacArthur and Levins 1967, Pianka 1981, Abrams 1983). Competition may not be active during long periods of relatively high resource availability, but episodes of low availability may significantly influence community structure (Wiens 1977).

My goals in this chapter are to explore patterns of coexistence among ant-following-guild members and to determine how resource use may differ among species. I focus on species-specific utilization of two army ants (*E. burchelli* and *L. praedator*) and on the dynamics of abundance and home-range size. I explore the inter- and intraspecific dominance hierarchies among the five bird species, and discuss where and how interference competition is most intense. Yearly population densities of the five bird species and two ant species are also explored in depth, and investigated as an influence on competition among the birds. I introduce "foraging space" as a conceptual tool related to the quantity of antswarm resource available to foraging birds over a given area. Foraging space is calculated using average swarm width per ant species, estimated density of each ant species over the study area, and daily probability that an ant colony will forage. Because birds aggregate at the front of a swarm, across its width, that calculation accurately reflects the average amount of "space" available to foraging birds per 100 ha. I compare predicted use of available foraging space, based on bird population density and antswarm density, with observed spacing behavior of individual birds at swarms, over three field seasons. Individual birds may need some minimum amount of space along a swarm's width to forage effectively among competitors. On a smaller spatial scale, I examine spacing behavior of birds along a swarm's width to better understand whether width represents a limiting resource for obligate ant-following birds.

To understand limitations on abundance of obligate ant-following birds, it was necessary to determine some basic yet unknown ecological parameters for the bird and army ant study species: (1) What is the colony density of *E. burchelli* at Cocha Cashu? (2) What is the density of active aboveground swarms of *L. praedator*? (3) What is the yearly adult population density of each of the five bird species at Cocha Cashu? (4) How is the dominance hierarchy structured among the five obligate ant-following bird species? And (5) what is the yearly mean home-range size for each of the five bird species in the study area? With estimates of those parameters, I can explore relationships among antswarm availability, competition between bird species, and differentiation in patterns of use of the army ant resource. My ultimate goal in this chapter is to determine how the five species of obligate ant-followers differ in their use of the available foraging resource, and how those differences potentially contribute to species coexistence. Using results from the basic ecological questions above, I explore the following questions:

I. Does body size or relative competitive ability among species correlate with home-range size?

II. Do obligate ant-followers minimize home-range size to fit the minimum number of reliable food resources needed for daily foraging?

III. Do obligate ant-following species

differentially utilize *E. burchelli* and *L. praedator* swarms and, if so, in what proportions?

IV. Are rates of agonistic interactions between birds different at *L. praedator* versus *E. burchelli* swarms?

V. Does body size or competitive ability correlate with space utilization at a swarm front?

VI. Judging from overall obligate ant-follower densities, are birds maximizing spatial utilization of available *E. burchelli* and *L. praedator* swarms?

VII. Can space at swarms be a limiting resource for obligate ant-followers?

Methods

Field Methods

Approximately 17 months of daily field observation of army ants and ant-following birds was carried out on a 161- to 277-ha study plot from September 1997 to April 2002. Almost all work was conducted in the rainy season, which coincides with the breeding season of obligate ant-following birds. It remains unknown whether behavior of the five obligate ant-followers changes in the dry season, when insect availability and foraging opportunities with *L. praedator* ants may decrease (Willis 1967, Levings and Windsor 1982).

Data collection was carried out by crews of two to five individuals per field season. Duration of the field season varied from year to year. In 1997, the study plot was selected and birds were intensively mist-netted and banded from September through November. In 1998 and 1999, the season lasted from September through December, with banding supplemented by antswarm observation and radiotelemetry. We continued that work from October 2000 through February 2001. In 2002, I returned for a month in March and April, mainly to gather data on adult survival rates.

Army ant field methods.—On arrival at the study site each year, my assistants and I (yearly total of four to five researchers), over the course of one to two weeks, walked trails throughout the study plot and located all *E. burchelli* colonies within the area. Trails were monitored throughout each field season, and new colonies were found as they entered the study plot. Each colony that was found was monitored and followed daily from bivouac to bivouac until it either left the study area or was lost by us, or until the field season ended. Loss of a colony generally resulted from the colony's movement into very thick, *Heliconia*-dominated herbaceous swamps, where tracking was difficult. Army ant tracking was not carried out in 1997. Ant colonies can move their bivouacs >150 m per night during the nomadic phase (S. K. Willson unpubl. data), and the best method for tracking a colony day-to-day is to follow its emigration column from one bivouac to the next. *Eciton burchelli* colonies at Cocha Cashu generally began moving their bivouac by late afternoon, and it was almost always possible to follow the moving column of ants to the new bivouac before nightfall.

Six days each week, we observed all foraging *E. burchelli* colonies for bird activity. We carried out observations for ~60 min per colony before noon, when army ant foraging is at its peak (Schneirla 1971). Observers stayed longer at large swarms that held more birds, to assure that all obligate ant-followers present had been accounted for. My assistants and I often monitored >10 colonies at a time, but with four researchers and the probability that some colonies were statary and not foraging on a given day, we were able to accomplish all foraging observations before noon each day. At the end of an observation, we recorded data on antswarm characters, including width of the swarm front. We checked all statary colonies daily for activity, and carried out an observation if the ants were foraging. We mapped all foraging fronts and bivouacs to the Cocha Cashu trail system, using estimates of degrees and meters from the nearest trail-marker. Observations were not carried out on Sundays, but all colonies that were likely to move to a new bivouac on Sunday night—nomadic colonies and those at the end of the statary phase—were checked that afternoon and followed to their new bivouac if necessary.

We found active aboveground swarms of *L. praedator* either opportunistically or while tracking birds with radiotelemetry. *Labidus praedator* swarms do not display the dendritic form characteristic of *E. burchelli* (see Chapter 2); rather, they form a dense "carpet" of ants. Measurements included $width_1$ and $width_2$, which we took perpendicularly after observing any ongoing bird activity in the same manner as at *E. burchelli* swarms.

Army ant densities.—We calculated average density of *E. burchelli* colonies at Cocha Cashu from known density measurements of bivouacs over a 50- or 100-ha area for 1998, 1999, and 2000–2001. We included a colony in weekly counts if its bivouac was located in the sampled area. I calculated a mean density from weekly counts over a field season. I calculated a core density, rather than the density of the entire plot, because plot size varied year-to-year and sampling was less intensive on the edges of the plot. Because of our intensive daily sampling in the core area, my assistants and I were aware of all *E. burchelli* colonies there, and mean densities from each season are based on actual counts of colony bivouacs rather than extrapolations or estimates (Franks 1982a). To account for time spent finding all army ant colonies on the plot, I omitted the first three weeks of each field season's *E. burchelli* colony counts.

In the 2000–2001 and 2002 seasons, I estimated density of active aboveground *L. praedator* swarms at Cocha Cashu. Unlike *E. burchelli*, *L. praedator* is subterranean as well as terrestrial, and ant-following birds must search for active aboveground swarms rather than tracking known colonies. I developed the following technique to estimate *L. praedator* density as available to ant-following birds. All researchers on the project recorded their own distance walked on trails in the study plot to the nearest 25 m and recorded the number of *L. praedator* swarms encountered on trails. We imposed a "15-minute rule" to mimic the movement speed of an average *E. burchelli* swarm front (because *L. praedator* movement speed has not been measured). If a trail was walked ≥15 min after it had previously been walked, that distance was included in the day's distance, to allow a hypothetical *L. praedator* front to move across a trail without being counted twice. *Labidus praedator* swarm movements are more *S*-shaped than linear; I assumed that swarm movement would still be forward rather than circular (that assumption seems appropriate, from personal observation of *L. praedator* swarms). My assistants and I coordinated our times and movements so that trails near camp would not be counted multiple times in a day. Total distances walked, per person per day, were summed to get a total distance (m) walked per week. The equation

$$\frac{\text{total } L.\ praedator\text{/wk}}{\text{total m/wk} * 3\text{ m}} * 1{,}000{,}000 = \frac{\text{est. \# swarms}}{100\text{ ha}} \quad (3.1)$$

describes the method for calculating total number of active *L. praedator* swarms per week. I multiplied distances by 3 to get an area (m^2) covered per week. Field trails were generally 2 m wide, and the area within 1.5 m of the center of the trail in either direction (total width 3 m) was easy to check for antswarm activity while observers walked trails at a normal pace. I calculated estimates of the number of active *L. praedator* swarms per 100 ha (1,000,000 m^2) each week of each field season and then averaged to get an overall mean for the season. Because the 2002 field season covered less than four weeks, all data were averaged to get one estimate for that season.

The technique described above is based on the assumption that trails throughout the study plot are a fair representation of the study plot as a whole. That assumption seems reasonable, because trails were cut in grid form to provide access to all habitats in the vicinity of the Cocha Cashu Biological Station (see Appendix 1 for a map of the trail system). *Labidus praedator* swarms were sometimes seen moving down trails, which would suggest that the trail was influencing movement direction. However, swarms larger than the trail width did not constrain their width to fit onto a trail, but spilled over the sides of the trail into surrounding vegetation, and we regularly observed swarms passing obliquely or perpendicularly over trails.

Avian field methods.—Mist-netting and banding methodologies follow the guidelines of the North American Banding Council (2001a, b, c). Mist-netting was carried out opportunistically, with lines of 10–20 nets placed near known *E. burchelli* bivouacs that were close to trails. Because the largest number of obligate ant-following birds are caught while ants are passing under nets, we strategically placed mist-nets to take advantage of army ant colonies moving through an area. We placed nets on trails in the afternoon, near the foraging path or bivouac of an army ant colony. At dawn the next day, we opened the nets and caught birds either as they flew to the bivouac to check its activity (Swartz 2001) or later as they passed with the army ants through the mist-net line.

Mist-nets were generally open from dawn until noon, though some late-afternoon netting was done. We outfitted captured birds with two individual color bands and one numbered metal band. I used standard techniques for morphological measurements: culmen length was measured from anterior edge of nares to bill tip, culmen width and depth were measured at anterior edge of nares, and wing chord was unflattened (North American Banding Council 2001a, b, c). Monomorphic species (*D. merula* and *P. nigromaculata*) and *R. melanosticta* juveniles were sexed using standard molecular gender-assignment techniques (Griffiths et al. 1998). I drew blood (30–50 μL) by brachial venipuncture. Extracted DNA was amplified using polymerase chain reaction (PCR), and PCR products were separated on a gel by electrophoresis to display sex-specific banding patterns.

During antswarm observations (described above), one or two observers positioned themselves ≥10 m from the sides of the swarm front, recorded all individual birds present, and took notes on behavior and interactions between individuals. We determined dominance within and between species by quantifying the number of times individuals displaced the perch site of other individuals during the observation period. I calculated percentages of perch or ground displacements that each species exhibited intra- and interspecifically to determine dominance relations and a dominance hierarchy among the five obligate ant-following bird species.

Beginning in 1998, we outfitted focal birds that were caught in mist-nets with radiotransmitters to monitor activity away from *E. burchelli* swarm fronts. We used Holohil BD-2 1.3-g transmitters, which weighed <5% of the bird's body weight and had a battery life of ~60 days. We evaluated three attachment methods: gluing to the bird's back, wing harness, and leg harness (Raim 1978, Rappole and Tipton 1991, Thompson 1994). Gluing to the back (using eyelash cement or "superglue") always failed within days, and wing harnesses were often quickly removed by the birds. Rappole and Tipton's (1991) leg-harness method worked well, and very few birds lost their radiotransmitters. In fact, three individuals that were not recaught by the end of the 2001 field season were still carrying transmitters 14 months later, in April 2002. One *M. fortis* male was recaught then and the transmitter removed; his skin showed no sign of abcess or irritation, though the cord was covered with shed skin around his upper legs. That skin came off easily, and the bird appeared healthy and unharmed by having carried the transmitter for a prolonged period. The leg-harness method also seemed to cause the birds the least stress, because placement was carried out quickly and did not irritate their skin. We followed birds with active radiotransmitters for 2- to 4-h blocks before and after noon; we collected data on their activity patterns, including use of nesting sites, roost sites, and *L. praedator* swarms. When birds were not directly observed (i.e. during roosting), we used triangulation from three points along a trail separated by ~50 m each (depending on distance from observer to bird). I later calculated locations of triangulated sites using the computer program LOCATE II (Nams 2000).

Statistical Methods

Avian population-density estimation.—I estimated adult population densities for each bird species each year, using Bowden's model estimation (Bowden 1993) in the program NOREMARK (White 1996). From that information, I calculated number of individuals per 100 ha and 95% confidence intervals for each species. The Bowden model estimation uses data from (1) number of banded birds, (2) number of sightings of banded but unidentified birds, and (3) number of sightings of unbanded birds over a season to estimate total adult bird density per species in the study area, plus 95% confidence intervals. That density is not a measure of "territorial" birds per area, because it includes unsettled adult floaters as well as adults with established home ranges. I treated individuals banded within the last two weeks of each field season as unbanded to accurately estimate the total adult population.

Home-range methods.—I calculated home ranges of birds using the program ANIMAL MOVEMENTS (Hooge and Eichenlaub 1997) in ARCVIEW. I present yearly data for kernel home ranges with sample sizes per individual per season of ≥20 points. Seaman et al. (1999) recommend that researchers obtain a minimum of 30 points, because kernel home-range estimates will overestimate home-range size at small sample sizes. Having >30 observations of few individuals per year, I was unable to use that cut-off; therefore, home ranges may be biased upward for some individuals. I used least-squares cross-validation to select the smoothing parameter of the fixed kernel (Seaman and Powell 1996), and report 95%-use areas. I included individual location points from sightings at swarms, *E. burchelli* bivouacs, mist-net records, and roost sites. Unlike generalist insectivores, which may have preferential feeding sites within territories, obligate ant-followers move through their home range following local army ant colonies and do not have designated foraging sites within home ranges. Observations of individual birds at swarms may bias the kernel method downward by allowing it to place separate contours around high-use points within a home-range area, thereby minimizing the 95%-use area. That situation could arise if a bird was observed at two spatially separated *E. burchelli* colonies but also, without an observer's knowledge, utilized an *L. praedator* swarm located between the two *E. burchelli* colonies. In that scenario, the bird may have utilized the area between the two contours of use at the *E. burchelli* swarms with the same frequency as within the contours, but was not detected there. I did not calculate minimum convex polygon (MCP) home ranges, which include all areas between points as part of a home range, because that method generally needs ≥150 points for accuracy (Seaman and Powell 1996).

I used a two-way analysis of variance (ANOVA) to test whether home-range size varied across species by year. I also used one-way ANOVA tests on home-range data from each species to examine whether there were significant differences in home-range size across years. Models did not include covariance to account for individuals observed in more than one year, because the number of those individuals per species was low.

Methods for Specific Questions

I. Does body size or relative competitive ability among species correlate with home-range size?—Home-range or territory size is the result of multiple factors, mostly related to population and resource density, defendability, and patchiness (Brown 1964). Agonistic interactions within (Willis 1967) and between species (Robinson and Terborgh 1995) also contribute to a bird's home-range size. Obligate ant-followers have been characterized in the literature as holding overlapping, non-exclusive home ranges rather than defended territories (Willis 1967, 1973; Willis and Oniki 1978). If territorial defense occurs, it is only near the nest site, as noted by Willis (1967). The five coexisting species at Cocha Cashu provide an opportunity to test competing hypotheses about home-range size. Density of army ant colonies may limit minimal home-range size, but relative mass of bird species or relative competitive ability (or both) may also be influential.

I proposed two competing hypotheses. Energetic requirements suggest that, all else being equal, larger animals should have larger home ranges than smaller animals (Schoener 1968). If body mass and energy requirements determine home-range size for obligate ant-following birds, we can predict that the largest species will have the largest home-range size and that home-range size will decrease as bird-species mass decreases. This hypothesis predicts a large, approximately equal home-range size for *D. merula*, *M. fortis*, and *P. nigromaculata*, with a progression to smaller home-range size in *R. melanosticta* and *G. salvini*. An important assumption of this prediction is that group size (solitary individuals, pairs, or family groups) will not influence home-range size. For the purposes of this prediction, I assumed that one antswarm could adequately feed any of the above group sizes. That assumption seemed reasonable, given that the yearly mean number of birds at *E. burchelli* swarms ranged from 5.7 to 8.9 individuals, and the lowest mean for number of birds at an *L. praedator* swarm was 3.4 (see bird numbers in Table 3.13).

Alternatively, I hypothesized that the dominance hierarchy among species influences home-range size. Dominant species may gather at highly productive swarms, denying subordinate species access to them. If subordinates need to sample more swarms to find sites at which to forage effectively, we can predict that they will have larger home-range sizes than dominant species. This hypothesis predicts that the small, subordinate species *R. melanosticta* and *G. salvini* have the largest average home-range sizes.

II. Do obligate ant-followers minimize home-range size to fit the minimum number of reliable food resources needed for daily foraging?—In addition to the hypotheses above, density of army ant colonies may be a major factor determining home-range size in obligate ant-following birds. The birds may attempt to minimize home-range size by tracking just enough army ant colonies to ensure daily foraging opportunities; that minimum may depend on body mass, competitive ability, or both. I assumed that a bird must forage each day, and modeled a predicted minimal home-range size using estimated army ant densities from Cocha Cashu and information on army ant foraging probabilities per day. I compared results of the model with results observed for each species to gain insight into species-specific ecological strategies and to test the predictions of body size and interspecific dominance on home-range size. First, I modeled the number of *E. burchelli* colonies a bird must keep track of per day to ensure daily foraging opportunities. That model assumed that obligate ant-following birds generally rely on *E. burchelli* for foraging opportunities, and use *L. praedator* only as a secondary source (Willis and Oniki 1978, Skutch 1996). I based that assumption on differences in reliability of the ant species. Obligate ant-followers monitor the mobile bivouacs of *E. burchelli* across space and time, but are unable to track colony movements of *L. praedator* ants, because that species does not have conspicuous, aboveground bivouacs.

A full cycle for *E. burchelli* lasts 35 days: 14 days in the nomadic phase and 21 days in the statary phase (Hölldobler and Wilson 1990). The nomadic phase—during which ants forage all day, every day, and move their bivouac almost every night—offers reliable food resources for birds. During the statary phase, however, the colony remains in its bivouac and does not forage every day. Using Franks' (1982b) estimate of 13 foraging days per statary phase (62% of statary days), I estimated that an *E. burchelli* colony has a 77.1% probability of foraging on any given day. The function

$$F_{eciton} = 1 - (1 - .77)^n \qquad (3.2)$$

describes the probability that at least one *E. burchelli* colony is foraging within a defined area, where n = the number of colonies in an area. For example, the probability that at least one *E. burchelli* colony is foraging on a given day is 77.1% for one colony, 94.7% for two colonies, and 98.8% for three colonies. The probability equations asymptote at approximately three *E. burchelli* colonies, so beyond that number it is likely unprofitable to follow more colonies (Fig. 3.1).

The function

$$F_{alternative} = 1.0 - F_{eciton} \qquad (3.3)$$

predicts the percentage of days an obligate ant-follower must rely on *L. praedator* swarms because no *E. burchelli* colonies are foraging in their home-range area. For example, if a home range encompasses two *E. burchelli* colonies, with a 94.7% chance that one is foraging on a given day, a bird will need to find an alternative food source (i.e. an *L. praedator* swarm) on 5.3% of days.

The factors that determine how many *E. burchelli* colonies a forager should track also depend on the density of those colonies in the landscape. If colonies are a limiting resource, their distribution may determine the minimum home-range size of an obligate ant-follower. The equation

$$\text{home range}_{minimum} = \frac{\text{\# colonies bird needs to track}}{(\text{colony density/100 ha})} \qquad (3.4)$$

varies depending on the number of colonies a bird simultaneously follows and the density of *E. burchelli* colonies in the area. For example, if colony density = 3.2 per 100 ha, as it does on Barro Colorado Island, Panama (Franks 1982b), an individual would need an average home range of ≥90.9 ha to encompass three colonies. If a 94.7% chance of foraging per day (or two colonies) is sufficient for a bird at that ant density, we can predict a minimum daily home-range size of 60.6 ha for obligate ant-followers on Barro Colorado. These percentages provide general daily probabilities of swarm availability to foraging birds. Given that

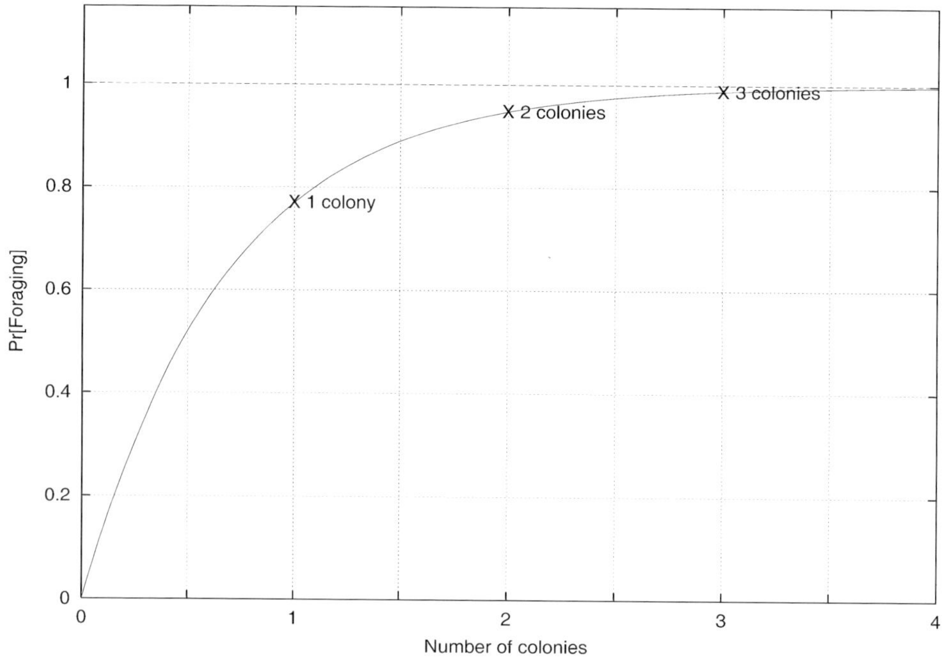

FIG. 3.1. The probability that ≥1 *E. burchelli* army ant colony is foraging, given the number of colonies in an area, asymptotes at ~3 colonies. Probability values are: 1 colony (77.1%), 2 colonies (94.7%), and 3 colonies (98.8%).

statary ant-colonies may not swarm all day (S. K. Willson pers. obs.), 77.1% may be too high an hourly probability of foraging for any given bird. However, for the purpose of the present analysis, the general probabilities should reflect reality for any given obligate ant-follower.

III. Do obligate ant-following species differentially utilize Eciton burchelli *and* Labidus praedator *swarms and, if so, in what proportions?*—I analyzed relative abundances of each bird species at *L. praedator* and *E. burchelli* swarms to determine whether they used the two species of army ants differently. Only data from the 2000–2001 field season were used, because the number of *L. praedator* swarm observations in previous years was small. I calculated total number of sightings of each bird species with each species of ant. I was unable to look for heterogeneity across the two ant species independently for each bird species, because our observations of the antswarms depended on finding the ants. We were consistently able to find *E. burchelli* swarms, because we followed the colonies every day; but finding an *L. praedator* swarm depended on following radiotracked birds, hearing bird vocalizations at active swarms, or occasionally happening upon an active swarm. Therefore, measurements of bird-species occurrence at *L. praedator* swarms are biased by the method of locating those swarms. Because *G. salvini* was the most evenly distributed of all bird species between swarms of the two ant species, I used them to correct for statistical sampling bias. All bird species were followed with radiotelemetry in approximately equal amounts.

I calculated relative percentage of abundance by dividing number of sightings of a bird species by total sightings of all obligate ant-followers, for each ant species. If all bird species proportioned their use similarly between *E. burchelli* and *L. praedator* swarms, we would expect that each bird species would make up the same relative percentage with each ant species. If any species varied from that pattern, we should see a corresponding shift in relative abundances of all other species. For example, if *D. merula* used *E. burchelli* swarms more than *L. praedator* swarms, relative percentage of *D. merula* should be higher with *E. burchelli* than with *L. praedator*, and all other bird species should have inflated relative abundances at *L. praedator* swarms to account for the absence of *D. merula*. My analysis showed that *D. merula* indeed had lower relative abundance at *L. praedator* swarms compared with its presence at *E. burchelli* swarms, but there was no corresponding increase in three of the other four species. Because those three species showed consistent abundances, I considered them "normal" for the purpose of comparing the other two species (*D. merula* and *M. fortis*) against them. *Gymnopithys salvini* had the most consistent relative abundance with the two ant species (23.3% and 22.6%, respectively), and I calculated a frequency value for each bird species, normalized to the frequency of *G. salvini* sightings. I compared normalized abundances using chi-square tests (Snedecor and Cochran 1989). Calculations of normalized abundance allowed me to determine whether *D. merula* and *M. fortis* utilized the two ant species differently from the other three bird species. By normalizing to one bird species, I also verified the assumption that *G. salvini*, *P. nigromaculata*, and *R. melanosticta* were behaving similarly.

IV. Are rates of agonistic interactions between birds different at Labidus praedator *versus* Eciton burchelli *swarms?*—I calculated displacement rates per minute per obligate ant-follower at *E. burchelli* swarms for 1998, 1999, and 2000–2001 and at *L. praedator* swarms for 1999 and 2000–2001. I calculated a two-way ANOVA to explore relationships among displacement rates, ant species, and year. I calculated a one-way ANOVA to test whether rates of agonistic interaction differed between years at *E. burchelli* swarms. A second one-way ANOVA tested the relationship between rate of displacements per minute and number of birds per unit of swarm-width at *E. burchelli* swarms. In all cases, "sample size" refers to total number of swarm observations used in the analysis.

V. Does body size or competitive ability correlate with space utilization at a swarm front?—I used a linear regression (SAS Institute 1997) to estimate meters of space an individual of each species used on an *E. burchelli* swarm front in the years 1998, 1999, and 2000–2001. To estimate a "meters per bird" (mpbird) parameter for each bird species, I fitted the number of birds of each species at each swarm front to the equation

antswarm width = [mpbird (*Phleg*) * number of *Phleg*] + [mpbird (*M.fort*) + number of *M. fort*] + [mpbird (*D.mer*) * number of *D.mer*] + [mpbird (*Rheg*) * number of *Rheg*] + [mpbird (*G.sal*) * number of *G.sal*] (3.5)

I estimated amount of space each species was predicted to use along an *E. burchelli* swarm front on the basis of four hypotheses. Space use at a swarm front may be influenced by (1) body mass, with larger species using more space than smaller species; (2) dominance, with dominant species using more space than subordinate species; or (3) ability to maximize space by choosing a swarm with fewer competitors. Hypothesis (3) is specific to *D. merula*, which may be physiologically able to sample more swarms because of its longer wing-shape and better flying ability as compared with the antbirds (S. K. Willson pers. obs.). Lastly (4, null hypothesis), space use along a swarm front may not be affected by the above factors, and instead may be similar for all species. Predictions for each hypothesis depend on yearly conditions and population parameters, and were calculated for 1998, 1999, and 2000–2001. For hypothesis (1), the equation

$$\text{Predicted m/bird} = \text{avg. m/bird} * M_s/\tilde{A} \qquad (3.6)$$

(where M_s = mass of bird species and $\tilde{A}$ = weighted average of bird mass based on population parameters of five species) provides predicted space use by each bird species at a swarm front, based on mass. For hypothesis (2), I predicted a linear decrease in space use corresponding to a species' position in the dominance hierarchy, with *P. nigromaculata* using the most space each year. For hypothesis (3), I predicted that *D. merula* space use at a swarm front would be within the upper 95% confidence interval of the dominant yet same-sized species *P. nigromaculata*. In hypothesis (4), I compared results for each species against the mean amount of space available per bird at an antswarm. That estimate is from the "observed space available per bird at an *E. burchelli* swarm" (see below). Predicted trends were visually compared with results from the 1998 and 2000–2001 field seasons—years with high and low total densities of ant-following birds, respectively.

VI. Judging from overall obligate-ant-follower densities, are birds maximizing spatial utilization of available Eciton burchelli *and* Labidus praedator *swarms?*—I use the term "foraging space" to describe the amount of antswarm resource available to foraging birds over a given area; this concept allows estimation of the obligate ant-following bird population's maximization of the available army ant resources. I calculated estimates of foraging space at *E. burchelli* and *L. praedator* swarms for three field seasons: 1998, 1999, and 2000–2001. I estimated mean *E. burchelli* swarm width from field data taken at swarms in each season. Mean *L. praedator* swarm width was available from the 2000–2001 season only; that width was used in calculations for all years. Likewise, density of aboveground *L. praedator* swarms was estimated only in the 2000–2001 season, and that estimate was used for all years.

Foraging space for each army ant species per year was calculated using the average swarm width, the estimated colony (for *E. burchelli*) or swarm (for *L. praedator*) density per 100 ha, and a general probability that a colony foraged on a given day (Franks 1982b). Probability of foraging per day is 1.0 in *L. praedator*, because the density measurement specifically estimated active aboveground swarms, not colonies. I used the following calculations to generate the data presented in Table 3.13 (letters in parentheses correspond to letters in that table). I estimated total foraging space available (i) to birds per year by summing the estimates of foraging space for *E. burchelli* (d) and *L. praedator* (h) for a given year. Mean number of obligate ant-following birds per *E. burchelli* or *L. praedator* swarm was estimated for a given swarm when bird presence was >0 (i.e. swarms with no birds present were not included in the estimate). I estimated amount of foraging space available per bird if all birds used only *E. burchelli* swarms by dividing the total available *E. burchelli* foraging space per 100 ha (d) by the total obligate ant-following population per 100 ha (j). I performed the same calculations with *L. praedator* data to estimate foraging space available per bird if all birds used only *L. praedator* swarms (h/j). I estimated the observed spacing per bird at swarms by dividing the mean swarm width by the mean number of obligate ant-following birds per species of swarm, for each year (b/n, f/q). To determine the maximal space a bird could obtain at a swarm front if all obligate ant-followers maximized their use of the two ant species, I calculated a measure of yearly maximal foraging space. Total foraging space available per 100 ha (i) was divided by the total population of obligate ant-followers per 100 ha (j). Finally, I asked how well birds maximized their use of the swarm resources each year. I estimated the percentage of maximization of total swarm resources twice: first, by dividing the observed spacing at *E. burchelli* swarms (o) by the maximum potential foraging space per bird (m); and second, by dividing the observed spacing at *L. praedator* swarms (r) by the maximum potential foraging space per bird (m).

VII. Can space at swarms be a limiting resource for obligate ant-followers? Answering that question does not require specific methods, but involves a synthesis of data gathered while answering other questions.

Results

Army Ant Colony Density

Mean *E. burchelli* density during the study period was 4.9 colonies per 100 ha. In 1998, *E. burchelli* colony density per 100 ha averaged 4.4 colonies over a 10-week period (Table 3.1). In 1999, density averaged 5.2 colonies over 13 weeks; in 2000–2001, density averaged 5.0 colonies over 15 weeks. There were no significant differences in colony density among years.

I estimated 20.7 active aboveground *L. praedator* swarms per 100 ha for the 2000–2001 field season, which fell exclusively within the rainy season (Table 3.2). Density may be lower in the dry season, when *L. praedator* colonies do more foraging below ground and at night (Schneirla 1971, E. O. Willis pers. comm.). However, an estimate of 19.9 swarms per 100 ha over 25 days in March–April 2002 may indicate that there is less variability in *L. praedator* density than is generally assumed to occur between wet and dry seasons. That estimate coincides with the transition into the dry season, when rainfall was scarce. However, the ground may have been wet enough to affect subterranean *L. praedator* foraging. Because most of my observations were made during the rainy season, the estimate of 20.7 active swarms per 100 ha is used in all further calculations.

Adult Avian Population Densities

Total adult population density per 100 ha for the five obligate ant-following species varied between 45 and 86.6 individuals per year, and declined by ~25% each year (Table 3.3). However, individual species generally did not show a constant decline throughout the study period. Based on confidence intervals, estimated densities of *P. nigromaculata* and *R. melanosticta* significantly declined between 1998 and 1999, and *R. melanosticta* and *G. salvini* declined between 1999 and 2000. Each year, relative abundance per species varied with changes in population densities; and each year, a different species was relatively most common on the study plot: in 1998, *P. nigromaculata*; in 1999, *G. salvini*; and in 2000–2001, *D. merula*.

Dominance Relationships among Bird Species

For species that differ in size, mass largely determined a species' rank in the strict dominance hierarchy that exists among the antbirds (Willis and Oniki 1978). Social system, as well, appeared to affect hierarchical placement when species were, on average, equal in mass. In contrast to the antbirds, the placement of the one woodcreeper species (*D. merula*) was less clear. Although it had the highest mass of the five species (see Table 2.1), *D. merula* interacted most with *R. melanosticta*, which is third in the

Table 3.1. Colony density measurements of *Eciton burchelli* army ants at Cocha Cashu Biological Station, Peru, 1998–2001. A one-way ANOVA found no difference in means between years (F = 0.527, df = 2 and 35, P = 0.60).

Year	Estimate	SD	SE	Area surveyed	Period
1998	4.4 per 100 ha	2.27	0.76	50 ha	10 weeks
1999	5.2 per 100 ha	2.39	0.69	50 ha	13 weeks
2000–2001	5.0 per 100 ha	1.20	0.35	100 ha	15 weeks
Mean	4.9 per 100 ha				

TABLE 3.2. Density measurement of aboveground foraging swarms of *Labidus praedator* army ants at Cocha Cashu Biological Station, Peru, from October 2000 to February 2001 (n = 126 days) and from March to April 2002 (n = 25 days).

Month(s)	Estimate	SD	SE
October 2001	26.1/100 ha	–	–
November 2001	20.4/100 ha	–	–
December 2001	14.7/100 ha	–	–
January 2001	22.1/100 ha	–	–
February 2001	20.1/100 ha	–	–
October–February 2001 average	20.7/100 ha	11.03	2.60
March–April 2002	19.9/100 ha	–	–

antbird dominance hierarchy and averages 16 g smaller than *D. merula* (Table 3.4). The dominant species is *P. nigromaculata*, a 46.0-g (on average) bird that lives in family groups of two to five individuals and is extremely aggressive interspecifically, with 69% of recorded displacements being directed at other obligate ant-following species. *Myrmeciza fortis* is the second species in the dominance hierarchy. Although it averaged 46.5 g, 0.5 g larger than *P. nigromaculata*, it is subordinate to that species and directed only 2% of displacements toward it. Eighty percent of recorded displacements by *M. fortis* were aimed at the two smaller antbird species. *Myrmeciza fortis* is extremely intraspecifically intolerant, and more than one pair never foraged together at the same swarm. As an *M. fortis* individual approaches a swarm, it sings loudly, and will retreat when answered by a bird other than its mate. When two pairs found themselves at the same swarm, fighting always ensued until at least one pair left the area. *Myrmeciza fortis* was the only obligate species at Cocha Cashu that defended multipurpose territories (see below). *Rhegmatorhina melanosticta* is the third-smallest antbird (average 31.4 g) and the third antbird in the dominance hierarchy. Only 2% of displacements by *R. melanosticta* were directed at the two larger antbirds; unlike those species, it directed most aggression (52%) at conspecifics. However, 16% of displacements by *R. melanosticta* were of *D. merula* (which weighs an average 47.5 g)—the highest displacement rate toward *D. merula* by any of the four antbird species. *Dendrocincla merula*, on the other hand, was almost exclusively intraspecifically aggressive; only 10% of its displacements were of antbirds, whereas 90% were of conspecifics. *Dendrocincla merula* woodcreepers preferentially cling to large trunks above advancing swarms (Willis 1978) and are less apt to contest perch sites with the obligate antbirds. However, they regularly fly to the ground to catch fleeing arthropod prey and will perch on the ground among antbirds while foraging. At the bottom of the dominance hierarchy is *G. salvini*, which weighs 25.9 g on average; 90% of its displacements were of conspecifics. As with *R. melanosticta*, multiple pairs may gather at an antswarm, and same-sex aggression between pairs was high.

AVIAN HOME-RANGE SIZE

Home-range size varied widely across species and over time (Table 3.5). Mean home ranges across all years, using individuals with ≥20

TABLE 3.3. Estimated number of individuals per 100 ha for five obligate ant-followers in Amazonian Peru. Estimates were calculated using Bowden's model estimation in the program NOREMARK. Population data are from 1998, 1999, and 2000–2001. Size of study plot varied by year (162 ha in 1998, 277 ha in 1999, 271 ha in 2000–20001); CI = 95% confidence interval.

Species	Estimate per 100 ha	CI 100 ha	Relative abundance (%)
1998			
D. merula	19.8	16.0–24.1	22.9
P. nigromaculata	21.0	16.7–26.5	24.2
M. fortis	9.3	3.1–27.2	10.7
R. melanosticta	16.7	14.2–19.8	19.3
G. salvini	19.8	14.8–27.2	22.9
Total	86.6		
1999			
D. merula	15.2	13.4–17.7	22.9
P. nigromaculata	13.0	11.6–14.8	19.5
M. fortis	10.1	7.9–12.3	15.2
R. melanosticta	10.1	8.3–11.9	15.2
G. salvini	18.1	13.4–24.5	27.2
Total	66.5		
2000–2001			
D. merula	14.0	11.1–18.1	30.6
P. nigromaculata	10.7	8.9–13.3	23.4
M. fortis	8.5	5.9–11.8	18.6
R. melanosticta	4.8	3.3–6.6	10.5
G. salvini	7.7	6.6–9.2	16.9
Total	45.7		

TABLE 3.4. Rates of displacement among five obligate ant-following bird species in Amazonian Peru. Rows represent proportion of total displacements directed at each of the five species (in columns) by the aggressor species (first in row). Sample size (n) of total displacements per species is shown in the last column.

	P. nigromaculata (46.0 g)	*M. fortis* (46.5 g)	*D. merula* (47.5 g)	*R. melanosticta* (31.4 g)	*G. salvini* (25.9 g)	n
P. nigromaculata	0.32	0.05	0.09	0.45	0.10	267
M. fortis	0.02	0.11	0.07	0.53	0.27	115
D. merula	0.01	0.00	0.90	0.03	0.06	463
R. melanosticta	0.01	0.01	0.16	0.52	0.31	444
G. salvini	0.01	0.00	0.02	0.07	0.90	154
Total n = 1,443						

observations within a season, were: *D. merula* = 64.6 ha (n = 20, SD = 33.8); *P. nigromaculata* = 45.9 ha (n = 20, SD = 30.9); *M. fortis* = 15.4 ha (n = 4, SD = 1.1); *R. melanosticta* = 50.8 ha (n = 24, SD = 41.5); and *G. salvini* = 27.5 ha (n = 21, SD = 13.2). Those averages, however, mask the strong effect of year on home ranges (ANOVA: F = 12.56, df = 2 and 86, P < 0.0001). In 1998 and 2000–2001, home-range size was similar within species; but in 1999, all study species increased their average home-range sizes by 19–59 ha (Fig. 3.2; *M. fortis* home-range size was not estimated for 1999 because there were too few observations per individual). Because of unequal variances and large standard deviations in *D. merula* and *G. salvini* data, only *P. nigromaculata* and *R. melanosticta* estimates are statistically significant for an increase in 1999 (ANOVA, *P. nigromaculata*: F = 38.53, df = 2 and 17, P < 0.001; *R. melanosticta*: F = 3.14, df = 2 and 21, P = 0.06). For at least one year per antbird species, mean home-range estimates were below the minimum needed to ensure that one of two *E. burchelli* colonies would be foraging in a home-range area on a given day (minimum = 40.8 ha, at 94.7% daily foraging probability; see Equation 3.4). Those results suggest that foraging with *L. praedator* army ants may be an important component of foraging strategy in each of the four antbirds, but may have been less important for the woodcreeper *D. merula*.

DISCUSSION

ECOLOGICAL RELATIONSHIPS AMONG BIRD SPECIES

1. Does body size or relative competitive ability among species correlate with home-range size?—There was little support for the hypothesis that home-range size was dependent on body mass. The three large ant-followers of equal size (*P. nigromaculata, M. fortis*, and *D. merula*) held strikingly different mean home-range sizes (Fig. 3.2). (In 1999, mean home-range sizes of all species were greatly inflated, and no patterns between species were evident. See Chapter 4 for discussion of 1999.) Home ranges of *D. merula* averaged more than 3–4× as large as those of the equivalently sized *M. fortis*. Average *M. fortis* home-range size was consistently smaller, not only than those of the two species of equal mass, but than those of all obligate ant-followers, including *G. salvini*, whose mass is only half that of *M. fortis*. Only the species pair *R. melanosticta* and *G. salvini* showed a trend correlating smaller home-range size with a decrease in body mass.

TABLE 3.5. Estimates for 95% kernel home range (with sample size and SD) for adult obligate ant-following bird species in 1998, 1999, and 2000–2001. Home ranges were estimated using the ANIMAL MOVEMENTS program for ARCVIEW. Estimates include only individuals with ≥20 sightings per season.

Species	Mean home range (ha)	Total n	Overall SD	1998 (n)	1998 SD	1999 (n)	1999 SD	2000–2001 (n)	2000–2001 SD
P. nigromaculata	45.9	20	30.9	26.83 (6)	5.74	86.01 (5)	13.70	24.86 (9)	13.18
M. fortis	15.4	4	1.1	15.38 (1)	–	–	–	15.50 (3)	1.31
D. merula	64.6	20	33.8	49.42 (6)	26.84	77.78 (7)	44.51	66.59 (7)	24.56
R. melanosticta	50.8	24	41.5	38.00 (10)	20.10	75.63 (11)	52.80	29.26 (3)	14.10
G. salvini	27.5	21	13.2	20.21 (5)	4.37	39.32 (4)	25.83	22.85 (12)	6.10

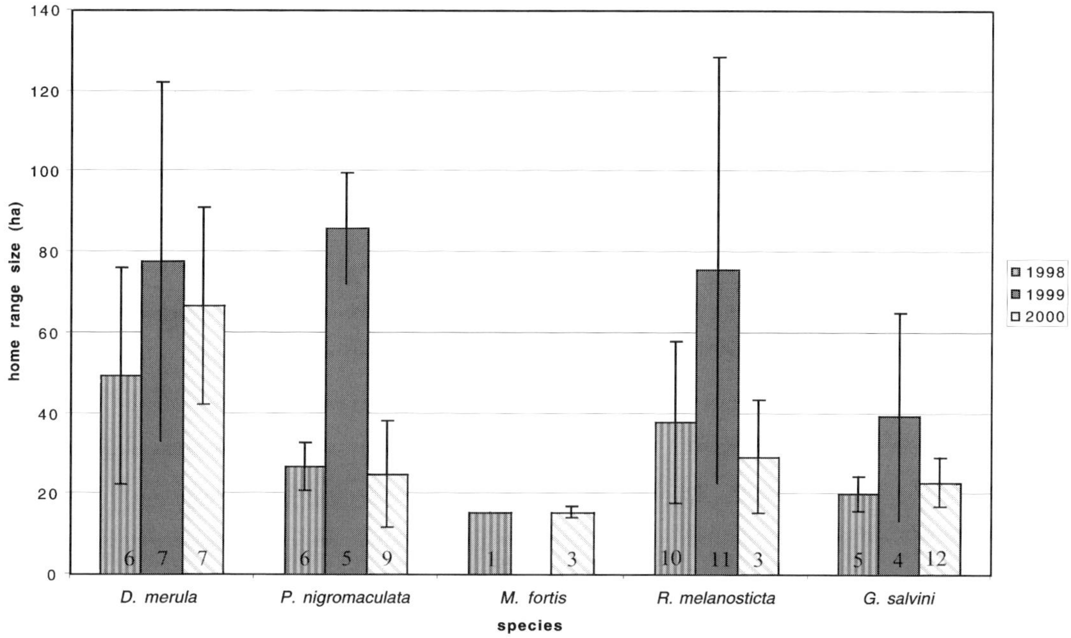

FIG. 3.2. Mean adult home-range size (95% kernel) for five species of obligate ant-following birds in Amazonian Peru over three years, with SD. Home ranges were estimated using the ANIMAL MOVEMENTS program for ARCVIEW. Sample sizes per year are shown within each bar; only individuals with ≥ 20 data points are included.

There was support for the hypothesis that placement in the interspecific dominance hierarchy affects home-range size. The greatest amount of interspecific aggression among the five bird species was directed at *R. melanosticta* (Table 3.4). The two dominant species, *P. nigromaculata* and *M. fortis*, displaced *R. melanosticta* more than any other species. In both 1998 and 2000–2001, *R. melanosticta* individuals held, on average, larger home ranges than *P. nigromaculata* and *M. fortis* individuals (Table 3.5).

Although home ranges of the subordinate species *G. salvini* were not larger than those of *R. melanosticta*, the former species received only 31% of the latter's aggression. *Rhegmatorhina melanosticta* is primarily intraspecifically aggressive, and may not have much influence on the movement patterns of *G. salvini*. *Gymnopithys salvini* consistently held larger home ranges than those of the largest antbird, *M. fortis*. In sum, there was little support for the hypothesis that body size affects relative home-range size and some support for the dominance-hierarchy hypothesis. However, mean home-range sizes for *D. merula*, which were consistently largest, and *M. fortis*, which were consistently smallest, were not explained by either hypothesis and will be further explored below.

II. Do birds minimize home-range size to fit the least number of food resources needed?—Animals use valuable energy stores when moving between patchy food resources (Charnov 1976, McNair 1982, Sutherland and Parker 1985). In the case of obligate ant-following species, which generally do not defend exclusive territories, food acquisition and position in the dominance hierarchy may act synchronously to determine home-range size. In terms of energetic demands, a minimum preferred home-range size may be one that holds a minimum number of army ant colonies providing some acceptable probability of foraging success. We have seen that the dominance hierarchy affects home-range size, but how do we predict the smallest home-range size an obligate ant-follower should have? Here, I ask whether colony density of *E. burchelli* army ants determines home-range size. That question is especially pertinent to the dominant species, *P. nigromaculata*, which is not constrained in home-range size by other guild members.

Using a mean of 4.9 *E. burchelli* colonies per 100 ha (Table 3.1) in the minimal-home-range equation (Eq. 3.4), we find that if an obligate ant-following bird depended solely on *E. burchelli* and tracked two colonies, it would require a minimum home range of 41 ha at Cocha Cashu to provide a 94.7% chance that one of the colonies was foraging on a given day. If the individual required the greater assurance of one of three colonies foraging on any given day, or a 98.8% chance of foraging, the minimum home range would be 61 ha. An alternative hypothesis is that obligate ant-followers do not need multiple *E. burchelli* colonies within their home range. At Cocha Cashu, a home range of 20 ha would provide an obligate ant-follower with a 77.1% probability that the colony is foraging on a given day.

In the 1998 and 2000–2001 seasons, the four antbird species all held average home-range sizes less than the minimum 41 ha needed to track two *E. burchelli* colonies (Fig. 3.2). Only *R. melanosticta* came close to the predicted minimum for two colonies, with an average home-range size of 38 ha in 1998. Average home-range size per year for the woodcreeper *D. merula* was greater than that needed to include three *E. burchelli* colonies in their home range in all years except 1998. As stated above, all home ranges in 1999 were greatly inflated; that inflation and its probable cause are further discussed below. These results suggest that, at least for the antbirds, *E. burchelli* colony density is not the only factor driving home-range size. Below, I examine whether home ranges are smaller than predicted because of bird use of *L. praedator* swarms.

III. Do bird species differentially utilize Labidus praedator *versus* Eciton burchelli *as a foraging resource?*—For each bird species, I determined the mean number of *E. burchelli* colonies and *L. praedator* swarms within the average home range per year. Using equations 3.1 and 3.2, I calculated the percentage of time a bird species should devote to each ant species if *E. burchelli* is the preferred resource (Table 3.6). By calculating the percentage of days that *E. burchelli* would not be available, I assumed that obligate ant-followers must use *L. praedator* swarms on those days. No obligate ant-following antbird was ever seen foraging away from army ants. It is unknown what percentage of the time the woodcreeper *D. merula* used white-lipped peccary herds, but I expect that percentage to be small, because those herds are spatially patchy and highly mobile at Cocha Cashu. The percentage of days that *E. burchelli* is unavailable to *D. merula* is presumably divided between foraging with *L. praedator* ants and with peccaries.

Using data from 2000–2001, I estimated that for *P. nigromaculata*, an average home range of 25 ha will encompass 1.2 *E. burchelli* colonies and 5.2 *L. praedator* swarms (Table 3.6). That means that on any given day, there is a 17% chance that no *E. burchelli* colony will be foraging. Thus, assuming that individual birds must forage every day, *P. nigromaculata* is predicted to have foraged with *L. praedator* on ≥17% of days in the 2000–2001 season. The smallest home range was consistently that of *M. fortis*. Using the same logic as above, that species was found to have had an average of 0.8 *E. burchelli* colonies within its home range and to have foraged with *L. praedator* on ≥29% of days in the 2000–2001 season.

Data from 1998 were generally similar to results from the 2000–2001 season (Table 3.6). In 1999, because home ranges of all species were much larger than in other years, expected percentage of days when no *E. burchelli* ant colonies were foraging was much lower. Only *G. salvini*, with a mean home-range size of 39.3 ha in 1999, was predicted to need a food resource other than *E. burchelli* on >1% of days. However, much *E. burchelli* foraging in 1999 was above ground in vegetation, because the study plot was inundated with rainwater—making those swarms less available to foraging birds (see below).

Because I was unable to monitor all *L. praedator* swarms in the study area as I did for *E. burchelli*, it is difficult to determine an average percentage of time that each ant species was used by the bird species. However, radio-tracked birds provide evidence that utilization of *L. praedator* is an important aspect of the study bird species' foraging ecologies (Table 3.7). Two *M. fortis* individuals were tracked extensively in 2000–2001 (a male and a female on separate territories). While being radiotracked, the male was observed foraging 13 times at temporally or spatially separated swarms, 54% of which were *L. praedator*. The female *M. fortis* was tracked to 19 swarms, 37% of which were *L. praedator*. Those numbers are greater than the 29% use of *L. praedator* predicted on the basis of average home-range size and preference for *E. burchelli* army ants. Predicted home-range sizes assume that a bird can move from one side of its home range to the other to effectively benefit from the number of *E. burchelli* swarms its home range contains, and that assumption is valid.

TABLE 3.6. Average home-range size per bird species per year allows a prediction of the number of foraging opportunities available each day with each of the two army ant species, *Eciton burchelli* (*E. b.*) and *Labidus praedator* (*L. p.*). The percentage of days when *E. burchelli* is unavailable provides a prediction of the percentage of days a species should attend *L. praedator* swarms (assuming that birds will preferentially use available *E. burchelli* swarms).

Species	Average home range (ha)	*E. b.* density per 100 ha	Number[a] expected with *E. b.*	Number[a] expected with *L. p.*	% of days *E. b.* available	% of days *E. b.* unavailable
			1998			
P. nigromaculata	26.83	4.4	1.2	5.6	82.9	17.1
M. fortis	15.38	4.4	0.7	3.2	64.3	35.7
D. merula	49.42	4.4	2.2	10.2	96.0	4.0
R. melanosticta	38.00	4.4	1.7	7.9	91.8	8.2
G. salvini	20.21	4.4	0.9	4.2	73.4	26.6
			1999			
P. nigromaculata	86.01	5.2	4.5	17.8	99.9	0.1
M. fortis	–	5.2	–	–	–	–
D. merula	77.78	5.2	4.0	16.1	99.7	0.3
R. melanosticta	75.63	5.2	3.9	15.7	99.7	0.3
G. salvini	39.32	5.2	2.0	8.1	94.7	5.3
			2000–2001			
P. nigromaculata	24.86	5.0	1.2	5.2	82.9	17.1
M. fortis	15.50	5.0	0.8	3.2	71.2	28.8
D. merula	66.59	5.0	3.3	13.8	99.2	0.8
R. melanosticta	29.26	5.0	1.5	6.1	89.0	11.0
G. salvini	22.85	5.0	1.1	4.7	80.1	19.9

[a] Of foraging opportunities.

TABLE 3.7. Frequency and relative use of *Eciton burchelli* and *Labidus praedator* swarms by individual radiotracked birds. All individuals listed had separate home ranges (none are mates or share the same home range).

Species	ID	Sex[a]	*n* (swarms)	% with *E. burchelli*	% with *L. praedator*	Year
P. nigromaculata	g-sr	U	5	60.0	40.0	2000–2001
P. nigromaculata	y-sr	M	20	80.0	20.0	2000–2001
P. nigromaculata	so-b	U	11	45.5	54.5	2000–2001
P. nigromaculata	oo-s	U	7	28.6	71.4	2000–2001
M. fortis	r-sg	M	14	46.2	53.8	2000–2001
M. fortis	g-sy	F	19	63.2	36.8	2000–2001
D. merula	wp-s	F	22	81.8	18.2	2000–2001
D. merula	s-pr	F	14	64.3	35.7	2000–2001
D. merula	rg-s	F	19	73.7	26.3	2000–2001
R. melanosticta	gp-s	F	56	76.8	23.2	2000–2001
R. melanosticta	r-sr	M	8	87.5	12.5	2000–2001
R. melanosticta	s-yr	M	20	80.0	20.0	1999
G. salvini	go-s	M	23	43.5	56.5	2000–2001

[a] U = unknown, M = male, F = female.

Individuals of all five bird species were observed attending *E. burchelli* swarms at the opposite extremes of their home ranges on consecutive days (S. K. Willson unpubl. data). All species displayed similar patterns of *L. praedator* swarm use (percentage of total swarms at which bird was observed while radiotracked), in that individual bird use of that ant species exceeded the prediction from home-range estimates (Tables 3.7 and 3.8). In *P. nigromaculata*, individual use of *L. praedator* swarms ranged from 18.2% to 35.7%, whereas predicted use from 2000–2001 home-range data was 17%. In *R. melanosticta*, individual use of *L. praedator* ranged from 12.5% to 23.2%, whereas predicted use was 11%. The one male *G. salvini* that was radiotracked used

TABLE 3.8. Relative abundance and normalized frequencies of five coexisting obligate ant-following bird species at *Labidus praedator* and *Eciton burchelli* antswarms in Amazonian Peru. Percentages were calculated from all adult sightings at each species' swarms over one field season in 2000–2001. Frequencies are normalized to those of *G. salvini* to correct for inherent differences between *E. burchelli* and *L. praedator* swarms. Abundance = percentage of obligate population at swarms of each ant species; frequency = number of individuals at a swarm for each individual of *G. salvini*; *n* = observed number of swarms of each ant species.

	L. praedator (*n* = 91)		*E. burchelli* (*n* = 247)				
	Abundance	Frequency	Abundance	Frequency	χ^2 value[a]	df	*P*-value[b]
P. nigromaculata	31.40%	1.35	29.60%	1.31	0.20	1	0.657
M. fortis	19.90%	0.85	12.00%	0.53	24.36	1	<0.001
D. merula	9.30%	0.40	20.80%	0.92	67.19	1	<0.001
R. melanosticta	16.10%	0.69	15.00%	0.67	0.10	1	0.757
G. salvini	23.30%	1.00	22.60%	1.00	NA	1	NA

[a] The χ^2 values were calculated using frequency normalized to total *G. salvini* sightings.
[b] $value_{crit} = 3.841$

L. praedator swarms 56.5% of the time; estimated use based on home-range size was 20%. Finally, three individual *D. merula* woodcreepers used *L. praedator* swarms 18% to 36% of the time, even though predicted use was <1%. In effect, those results emphasize the importance of *L. praedator* as a food resource for obligate ant-followers, and provide evidence that individual birds are choosing to forage opportunistically at spatially unpredictable *L. praedator* swarms rather than temporally and spatially reliable *E. burchelli* swarms. Below, I explore whether individual bird species had an army ant species preference. I also explore how rates of agonistic encounters differed between ant species and between years of high and low bird-density.

Data from 339 antswarms observed in the 2000–2001 field season were analyzed to determine relative abundance and normalized frequency of each bird species at *E. burchelli* (*n* = 247) and *L. praedator* (*n* = 91) swarms (Table 3.8). *Phlegopsis nigromaculata, G. salvini,* and *R. melanosticta* all used the two ant species in the same way, judging from normalized frequencies. More than those three antbird species, the woodcreeper *D. merula* favored *E. burchelli* (20.8% relative abundance at *E. burchelli* swarms versus 9.3% at *L. praedator* swarms). *Myrmeciza fortis* used *L. praedator* swarms more than the other birds (19.9% relative abundance at *L. praedator* and 12% at *E. burchelli*). Both *D. merula* and *M. fortis* showed statistically significant differences in ant-species use, compared with the other three bird species (χ^2: $P < 0.001$; Table 3.8).

IV. Are rates of agonistic interactions between birds different at Labidus praedator *versus* Eciton burchelli *swarms?*—Mean rates of agonistic interactions, as described by displacements of perch sites, varied between years and between ant species. Mean displacement rate per individual at *E. burchelli* swarms was highest in 1999 (x = 0.0054) and lowest in the 2000–2001 season (x = 0.0018) (Table 3.9). Rate of displacements per minute per obligate ant-follower was dependent on year ($P = 0.049$), but neither species of army ant nor species × year interaction was significant (1998 was not included because there were too few *L. praedator* samples in that year). Year was highly significant when rates of displacement were compared among only *E. burchelli* swarms in 1998, 1999, and 2000–2001 ($P < 0.001$). Confidence intervals for the three field seasons show very little overlap (Table 3.10). Not surprisingly, the relationship between rate of displacements per minute and number of birds per unit of swarm width at *E. burchelli* swarms was significant ($P < 0.001$). In contrast to 1998 and 1999, during the 2000–2001 season the rate of agonistic interaction depended on how many birds were present at a swarm (Table 3.11). The less space individual birds had along the front, the more they fought. When population density was higher during 1998 and 1999, the rate of agonistic interactions was higher, no matter how many birds were at a particular swarm front. An interpretation of those data is that the rate increased to the point where it ceased to be dependent on the birds immediately present at a given swarm. Rather than fighting for more foraging space at the swarm, the birds seemed to engage in general competition against any subordinate birds present, even

TABLE 3.9. Average displacement rate per minute per obligate ant-follower at *Eciton burchelli* and *Labidus praedator* swarms (in bold) differs by year and by ant species. "Avg. min" is the average amount of time a swarm was observed, "Avg. displ. obs.$^{-1}$" is the average number of displacements per observation, "Avg. no. obligates" is the average number of obligate ant-followers at a swarm, "Displ. min^{-1}" is the average rate of displacements per minute, and "Displ. min^{-1} obl.$^{-1}$" is the average rate of displacements per minute per obligate ant-follower present at a swarm.

	1998	1999		2000	
	E. burchelli	*E. burchelli*	*L. praedator*	*E. burchelli*	*L. praedator*
Total displacements	549	606	14	356	44
Avg. min	79	85	53	74	53
SD min	36.6	37.3	50.0	45.4	31.9
Avg. displ. obs.$^{-1}$	3.50	3.67	0.70	1.50	0.58
SD displ. obs.$^{-1}$	5.96	6.34	1.49	3.85	1.61
Avg. no. obligates	9.41	7.81	4.05	5.69	3.28
SD obligates	5.55	5.09	1.93	3.74	2.13
Displ. min^{-1}	0.0371	0.0392	0.0139	0.0148	0.0088
SD displ. min^{-1}	0.0479	0.0637	0.0298	0.0334	0.0257
Displ. min^{-1} obl.$^{-1}$	**0.0034**	**0.0054**	**0.0028**	**0.0018**	**0.0018**
SD displ. min^{-1} obl^{-1}	0.0041	0.0139	0.0068	0.0036	0.0048
Number of swarms (*n*)	157	166	20	238	76

TABLE 3.10. Displacements per minute per ant-following bird (D min^{-1} bird^{-1}) at *Eciton burchelli* swarms, with 95% confidence intervals (CI). Overall model using ANOVA weighted for unequal variances was significant ($F = 12.588$, df = 2 and 292, $P < 0.001$).

Year	*n*	D min^{-1} bird^{-1}	SD	95% CI
1998	157	0.003448	0.00414	0.002160–0.00473
1999	166	0.005398	0.01386	0.004150–0.00665
2000	238	0.001759	0.00364	0.000715–0.00280

if there was plenty of space at that particular swarm.

V. Does body size or competitive ability correlate with space use at a swarm front?—I tested four hypotheses regarding swarm-front utilization: (1) space use is directly related to bird species mass; (2) space use is related to rank in the dominance hierarchy; (3) woodcreeper space use is significantly higher than that of the antbird species, because its longer wings and stronger flying ability allow *D. merula* to sample multiple swarms with less energy expenditure; and (4) birds divide swarm-front space evenly among species.

Data from three seasons were examined (1998, 1999, 2000–2001), and a linear fit was possible for two of those seasons (1998 and 2000–2001). A linear fit was not biologically meaningful for 1999, because the fit resulted in negative estimates for some species; results for that year are not discussed here. For both the 1998 and 2000–2001 field seasons, number of birds of each species at an *E. burchelli* swarm front correlated well with that swarm's width ($r^2_{adj.} > 0.66$, $P < 0.001$ in each year). Those results indicate that individuals of each species used consistent amounts of space within each of the two years.

Estimated space use ranged from 0.51 to 1.0 m per species in 1998 and from 0.73 to 1.9 m per species in 2000–2001 (Fig. 3.3). In 1998, space use was similar among the three species of equal mass (*P. nigromaculata, M. fortis*, and *D. merula*), at 0.87, 1.0, and 0.99 m, respectively. Those results provide some support for hypothesis (1) (Table 3.12). The two smaller antbirds,

TABLE 3.11. Displacements per minute per bird for three years, grouped by birds per meter of antswarm front.

Birds m^{-1}	1998	*n*	1999	*n*	2000	*n*	All years	*n*
≤0.5	0.0012	21	0.0054	48	0.0008	104	0.0021	173
>0.5–1.0	0.0048	31	0.0047	57	0.0023	100	0.0034	188
>1.0–1.5	0.0041	39	0.0045	21	0.0025	33	0.0036	93
>1.5	0.0036	43	0.0049	28	0.0042	35	0.0041	106

R. melanosticta and *G. salvini*, used less space, at 0.51 and 0.69 m, respectively. In opposition to hypotheses (1) and (2), the smallest and most subordinate species, *G. salvini*, used 0.18 m more space than *R. melanosticta*. With 95% confidence intervals, all species overlapped in swarm-front use estimates, though *D. merula* and *R. melanosticta* overlapped only slightly.

In 2000–2001, all species used more space along swarm fronts than in 1998, and two of the four hypotheses were supported (Fig. 3.3 and Table 3.12). Space use along a front corresponded with body mass (hypothesis 1) in all five species. Woodcreeper *D. merula* used the most space along the front, 1.9 m—an estimate even greater than that predicted using the upper 95% confidence interval of the dominant antbird *P. nigromaculata*—which supports hypothesis (3). The two antbirds of equal size, *P. nigromaculata* and *M. fortis*, used similar amounts of space (1.25 and 1.32 m, respectively). Hypotheses for size and dominance hierarchy were supported for the two smallest antbirds, with *R. melanosticta* space-use estimated at 0.95 m and that of the subordinate *G. salvini* estimated at 0.73 m. Confidence intervals for *G. salvini* did not overlap with those for *P. nigromaculata* or *D. merula* and overlapped slightly with those for *M. fortis*. Confidence intervals for *D. merula* overlapped slightly with those for *M. fortis*, but were above those for all other bird species.

Increased front-space per species in 2000–2001 corresponded with a drop in both intra- and interspecific competitive pressures (see Table 3.9), because total density of ant-following birds was almost half what it had been in 1998 (Table 3.3). Higher support for space use based on body mass in 2000–2001 than in 1998 may be related to the population decline over that period. In 1998, all antswarms held the maximal number of birds observed over the study period, and birds may not have been able to maximize space use because of the sheer numbers of individuals per swarm. That also provides an explanation for why *D. merula*, in 1998, was unable to "choose" swarms that offered more space along the swarm front; all swarms were maximally attended in that year.

VI. Judging from overall obligate-ant-follower densities, are birds maximizing spatial utilization of available swarms?—With detailed information on densities of obligate ant-following birds, army ant densities and foraging dynamics, and swarm widths, it is possible to determine the average

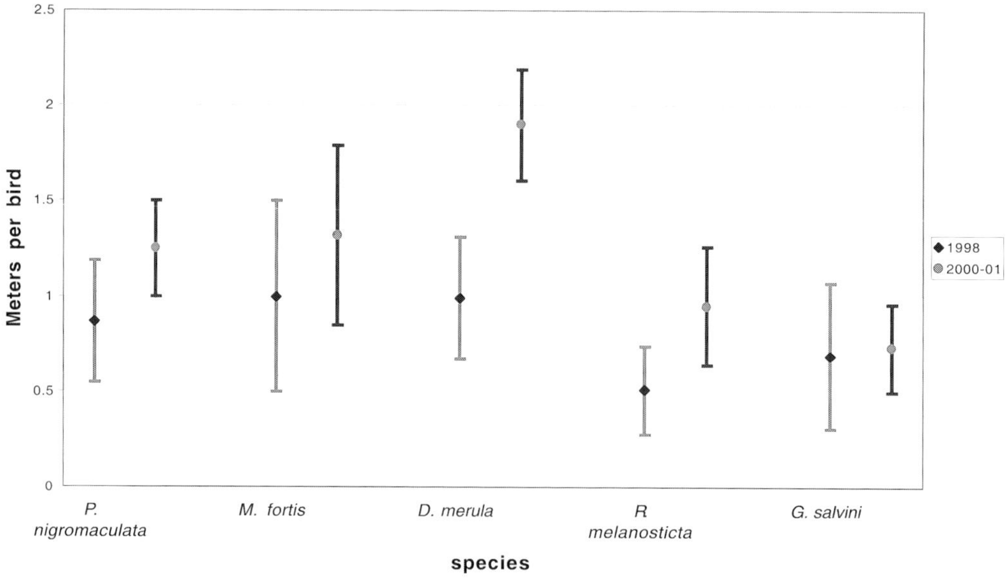

FIG. 3.3. Estimated meters of antswarm front used by individuals of five obligate ant-following bird species in Amazonian Peru in 1998 and 2000–2001, with 95% confidence intervals. Yearly estimates are from a linear regression of swarm width based on the number of birds of each species present at *E. burchelli* swarms (n = 138 swarms for 1998, 227 swarms for 2000–2001).

TABLE 3.12. Estimates of space use by birds along *Eciton burchelli* swarm fronts: predictions of four hypotheses and results for two field seasons. Hypotheses (1), (3), and (4) provide estimates of meters of space an individual bird of each species is predicted to use. Hypothesis (2) predicts a ranking of relative space use. Units of predictions and results are in meters (m) per bird and are provided for one field season of high bird-population density (1998) and one of low bird-population density (2000–2001). Predictions for hypothesis (3) are based on the upper 95% CI of the dominant bird species (*P. nigromaculata*). Predictions for hypothesis 4 are based on the average meters per bird observed across species in each year. No hypothesis was fully supported in 1998; hypotheses (1) and (3) were supported in 2000–2001.

	1998	2000–2001
	m bird^{-1}	m bird^{-1}
Predictions		
Hypothesis 1: Space use is related to species mass		
P. nigromaculata (46.0 g)	1.20	1.72
M. fortis (46.5 g)	1.22	1.74
D. merula (47.5 g)	1.24	1.77
R. melanosticta (31.4 g)	0.82	1.17
G. salvini (25.9 g)	0.68	0.97
P. nigromaculata = *M. fortis* = *D. merula* > *R. melanosticta* > *G. salvini*		
Hypothesis 2: Space use is directly related to position in dominance hierarchy		
P. nigromaculata > *M. fortis* > *D. merula* = *R. melanosticta* > *G. salvini*		
Hypothesis 3: *D. merula* space use is significantly greater than antbird space use		
D. merula	1.50	1.74
D. merula = upper 95% CI of *P. nigromaculata*		
Hypothesis 4: Birds divide swarm front evenly among species		
All species	1.02	1.58
P. nigromaculata = *M. fortis* = *D. merula* = *R. melanosticta* = *G. salvini*		
Results		
P. nigromaculata (46.0 g)	0.87	1.25
M. fortis (46.5 g)	1.00	1.32
D. merula (47.5 g)	0.99	1.90
R. melanosticta (31.4 g)	0.51	0.95
G. salvini (25.9 g)	0.69	0.73

daily foraging space available to the obligate ant-following population on the study site each year and to estimate overall bird use of that space. That estimate simplifies community dynamics in terms of both intra- and interspecific competition, by assuming that individual birds have knowledge of and access to all available swarms they encounter and that all swarms are equal in quality. In reality, despotic dominance both within and between species may affect antswarm choice. Use of all available swarm resources will theoretically follow an ideal preemptive distribution pattern (Pulliam and Danielson 1991) more than a pattern of ideal free distribution (Fretwell and Lucas 1970) and may affect the actual use of space along swarm fronts.

Estimated mean daily amount of foraging space available to the obligate ant-following bird population per 100 ha was 92.5 m in 1998, 104.5 m in 1999, and 96.4 m in 2000–2001 (Table 3.13). If ant-followers maximize spatial use of *L. praedator* and *E. burchelli* swarms, observed average spacing per bird at swarms should approach the maximal estimated amount of foraging space available per individual. If all obligate ant-following individuals used *E. burchelli* swarms, but never used those of *L. praedator*, a minimum amount of space would be available per individual at *E. burchelli* swarm fronts. For example, based on those totals and the population density of ant-following birds in 1998, individual birds could gain ≤1.07 m per individual if all birds were maximally using the swarm resources of *E. burchelli* and *L. praedator*, but ≤0.36 m if all birds foraged solely with *E. burchelli* (Table 3.13). Observed foraging space per individual at *E. burchelli* swarms was 1.02 m per individual, or 95.3% maximization.

TABLE 3.13. Comparison of observed space use by ant-following birds at antswarms. The theoretical maximum based on ant availability provides an estimate of how well the birds maximize their use of available foraging space (in bold). Estimates are given for *Eciton burchelli* (*E.b.*) and *Labidus praedator* (*L.p.*) swarms; density of *L. praedator* was measured in 2000–2001 only, and is assumed constant across years.

	1998	*n*	1999	*n*	2000-01	*n*
a. *E.b.* density per 100 ha	4.4	10 weeks	5.2	13 weeks	5.0	15 weeks
b. Average width of *E.b.* swarms (m)	9.1	102	10.7	112	9.0	196
c. Chance of foraging per day	77%	[a]	77%	[a]	77%	[a]
d. Space at *E.b.* per 100 ha [abc] (m)	30.8	–	42.8	–	34.7	–
e. *L.p.* density per 100 ha	–	–	–	–	20.7	15 weeks
f. Average width of *L.p.* swarms (m)	–	–	–	–	2.98	91
g. Chance of foraging per day	100%	–	100%	–	100%	–
h. Space at *L.p.* per 100 ha [efg] (m)	–	–	–	–	61.7	–
i. Total foraging space per 100 ha [d+h] (m)	92.5	–	104.5	–	96.4	–
j. Total birds per 100 ha	86.6	NA	66.5	NA	45.7	NA
k. Space per bird if all at *E.b.* [d/j] (m/bird)	0.36	–	0.64	–	0.77	–
l. Space per bird if all at *L.p.* [h/j] (m/bird)	0.71	–	0.93	–	1.35	–
m. Maximum potential spacing [i/j] (m/bird)	1.07	–	1.57	–	2.11	–
n. Average birds per *E.b.* swarm	8.9	102	7.8	112	5.7	196
o. Observed spacing at *E.b.* [b/n] (m/bird)	1.02	–	1.37	–	1.58	–
p. **Maximization of space at *E.b.*** [o/m] (%)	**95.3**	–	**87.3**	–	**74.9**	–
q. Average birds per *L.p.* swarm	5.3	6	4.1	27	3.4	91
r. Observed spacing at *L.p.* [f/q] (m/bird)	0.56	–	0.74	–	0.88	–
s. **Maximization of space at *L.p.*** [r/m] (%)	**52.3**	–	**47.1**	–	**41.7**	–

[a] (Franks 1982b)

In 1999, maximal foraging space was 1.57 m. If all birds foraged only with *E. burchelli*, foraging space would be reduced to 0.64 m. Observed spacing at *E. burchelli* swarms was 1.37 m, or 87.3% maximization of available resources. In the 2000–2001 season, birds obtained an average foraging space of 1.58 m per individual at *E. burchelli* swarms, or 74.9% maximization. If birds had not foraged with *L. praedator*, they would have obtained a maximum spacing of only 0.77 m per individual.

Actual spacing per bird at *L. praedator* swarms was always near or below half of what it might have been had the birds taken advantage of all antswarm resources available to them, and rates of maximization of *L. praedator* swarms ranged from 41.7% to 52.3%. That is based on the assumption that *L. praedator* density was the same in the other years as in 2000–2001. However, that assumption seems reasonable, given that I estimated 19.9 aboveground swarms per 100 ha for a two-month sample in 2002 (Table 3.2).

For all three years, the percentage of utilization of antswarm resources differed dramatically when spacing behavior at *L. praedator* swarms was calculated versus spacing behavior at *E. burchelli* swarms. Two main factors help explain why the estimate using *L. praedator* swarms is less useful than that obtained with the *E. burchelli* data. First, as explained earlier, *L. praedator* swarm use is dependent on the obligate ant-following birds' ability to find the swarms. Birds cannot efficiently use a resource if they do not know where it is, and birds only find *L. praedator* swarms opportunistically. Second, *L. praedator* swarms are often small enough that only one family group will find and forage at a particular swarm during an observation period. A family will have decreased agonistic interactions, and may use a swarm too small for multiple unrelated individuals to forage at together without aggression. Data from families at small *L. praedator* swarms will lower the observed foraging space for birds at *L. praedator* swarms and produce a lower estimate for maximization of foraging resources with that ant species.

Because *L. praedator* swarms cannot be followed by birds day-to-day with the precision that an *E. burchelli* colony can, utilization of the former species is not expected to approach the maximum possible, and percentage of use of total swarm resources is lowered. However, the estimate of swarm use in 1998 approaches complete utilization of swarm resources. The 1998 season had the highest density of obligate ant-followers on the study plot, the least number of *E. burchelli* colonies (4.4 per 100 ha), and the lowest amount of available foraging space with

E. burchelli (Table 3.13). It is possible that birds were more reliant on *L. praedator* in 1998 to avoid agonistic encounters at the fewer *E. burchelli* swarms. If birds tracked areas where *L. praedator* swarms had been previously encountered rather than locating those swarms opportunistically, that could raise the percentage of use of resources for that year.

VII. Can space be a limiting resource for obligate ant-followers?—Obligate ant-following birds are dependent on the food resource provided by foraging swarms. As army ants move over the forest floor, arthropods are continually flushed ahead of the oncoming ants. Paradoxically, that "superabundant" prey has allowed the ant-followers to evolve as highly specialized foragers, but it is also what keeps their populations in check. Competition is density-dependent, and although antswarms appear to be an unlimited resource, space along a swarm front is mediated through intense interference competition (Tables 3.9–3.11). Ability to capitalize on prey availability rests on a species' requirement of a particular amount of foraging space along the front. Space requirements were shaped by both body mass and density-dependent effects of competition, and space use per species increased in the year with the least competitive pressure and lowest avian population density (Fig. 3.3). Competition at swarms may be made more intense by "bet-hedging" resulting from birds' uncertainty about future availability of food. Particularly in the rainy season, there is always the chance that a heavy downpour will begin, which may preclude ant-followers from foraging the rest of the day. If rain is heavy enough, army ants stop foraging and begin a mass movement returning to the bivouac (S. K. Willson pers. obs.). Obligate ant-followers cannot depend on availability of their food resource for immediate future use, and thus cannot afford to "wait" or "sit out" a foraging bout when competitors are already at a swarm front. Future uncertainty and the physiological need to eat every day may increase what competition already exists for the limiting resource of space along a swarm front.

Avian Home Ranges and Populations

Data presented here strongly suggest that *L. praedator* swarms play a key role in allowing obligate antbirds to decrease home-range size. Each of the four antbird species held average home ranges, in at least one year, that were below the minimum size needed to secure daily foraging opportunities with *E. burchelli* alone. Additionally, radiotelemetry of individual birds shows that use of *L. praedator* swarms exceeds the percentage expected if bird species foraged with that ant only when *E. burchelli* was unavailable.

No previous study has measured the ecological value of *L. praedator* to obligate ant-following birds. Willis (1966b, 1972) described their use by facultative and migrant birds as high, because of lack of competitors. Generalized accounts of birds and army ants in the secondary literature imply that *L. praedator* is a relatively unimportant, marginally used resource in obligate ant-followers. Data presented here demonstrate the contrary: although spatially and temporally unreliable, an unoccupied *L. praedator* swarm can offer foraging opportunities unencumbered by aggression from other obligates, and a superabundance of prey. Foraging *L. praedator* colonies seem common enough at Cocha Cashu that obligate ant-followers can greatly benefit from using the resource opportunistically. Additionally, averages for the four antbirds suggest that home-range size is minimized through use of *L. praedator* swarms.

Two of the study species, *M. fortis* and *D. merula*, deserve special mention here. Home-range size of *M. fortis* was consistently smaller than those of other ant-following species. *Myrmeciza fortis* proved an indefatigable follower of *L. praedator* and, compared with other ant-followers, spent a significantly higher percentage of time with that species than with *E. burchelli* (Table 3.8). Density of available *L. praedator* swarms determined the small territories that *M. fortis* pairs defended. To the extent that there is any predictability in *L. praedator* foraging patterns, it ought to be most apparent to *M. fortis*, given that they spend the most time with *L. praedator* and hold the smallest territories. That they defend small, exclusive territories also suggests that *M. fortis* individuals should know their territories better than other ant-followers at Cocha Cashu. Data presented here seem to unravel some of the mystery of whether *M. fortis* obligately or facultatively follows army ants. To assume that *E. burchelli* is the preferred army ant resource of all obligate ant-followers might lead to the erroneous conclusion that *M. fortis* is not obligately tied to army ants.

The woodcreeper *D. merula* consistently had home ranges far larger than those of the four antbird species. It seems likely that flying ability may at least partially account for that. Antbirds typically fly between swarms by making short horizontal "hops" of 3–10 m from sapling to sapling (S. K. Willson pers. obs.). *Dendrocincla merula* individuals, on the other hand, are powerful fliers and can rapidly move hundreds of meters, typically flying through the understory at a height of ~2 m (S. K. Willson pers. obs). In that way, they can sample multiple antswarms with ease, assess potential competitors and mates, and search for peccary herds. Cocha Cashu Biological Station is unique among Neotropical research areas in that it has a large, healthy population of white-lipped peccaries (Terborgh et al. 1984)—a globally threatened species listed by the Convention on International Trade in Endangered Species (CITES), having been hunted to extinction in many areas of the Neotropics (Emmons 1990). At Cocha Cashu, I determined that *D. merula* woodcreepers supplement their foraging at antswarms by foraging with peccaries. More than obligate ant-followers, they are obligate followers of "beaters," and peccary herds provide that service as they move through an area, rooting the soil and uncovering arthropods in the leaf litter. One *D. merula* individual was seen perched on the hindquarters of a peccary (T. Baggalay pers. comm.), and they have also been observed perching on trees above herds. Additionally, >50% of *D.merula* mist-net captures in the present study smelled strongly of white-lipped peccary. Given that white-lipped peccaries are absent or greatly reduced in numbers over much of their former range, this interesting species interaction necessarily does not occur in most *D. merula* populations. It is unknown how the extirpation of white-lipped peccaries has affected *D. merula* abundance in other regions. However, the *D. merula* population north of Manaus, Brazil, is estimated at 3.6 birds per 100 ha during peak (postbreeding) density (Marantz et al. 2003). During the present study, average adult *D. merula* density at Cocha Cashu was 5–6× higher, not including juveniles. Although white-lipped peccaries are listed as present north of Manaus, Brazil, low herd density—resulting from forest fragmentation and hunting pressures—may have negatively affected historical density of *D. merula* in central Amazonian Brazil. That hypothesis is based on the assumption that white-lipped peccaries played a prominent role in the foraging ecology of *D. merula* in all regions where their historical ranges overlapped.

In a study of bird community structure at Cocha Cashu, Terborgh et al. (1990) estimated home ranges for four of the five species investigated here (*P. nigromaculata, M. fortis, R. melanosticta*, and *G. salvini*). Given the effort involved in that large-scale community study—carried out over three months of the dry season in 1982 and mainly using mist-net captures to create home-range sizes for those birds—sample sizes for some species estimates were necessarily small. My results refine Terborgh et al.'s (1990) estimated home ranges of >25 ha for *R. melanosticta* and *G. salvini*, but disagree with their results for *P. nigromaculata* and *M. fortis* (Table 3.14). Using sample sizes of just three and one, respectively, they estimated home-range size of *P. nigromaculata* at 14 ha, and that of *M. fortis* at >50 ha. Although some *P. nigromaculata* individuals had small home ranges, my average for the three years of data was 46 ha, with yearly averages ranging from 25 to 86 ha. The large home-range size calculated for *M. fortis* by Terborgh et al. (1990) may be an artifact of their sampling. Over three years of intensive observation of color-banded individuals, it is clear that mated *M. fortis* pairs defend small, intraspecifically exclusive territories. It is possible that the one individual spot-mapped in Terborgh et al.'s (1990) study had recently lost its mate and was searching over a large area for a new mate. In 1998, I observed a male *M. fortis* performing similar movements after his mate died. Eventually, he settled in a new area with a new female. Unfortunately, inaccurate estimates of extremely large home ranges in obligate ant-following birds are passed through the literature, perpetuating some of the misconceptions about these species (e.g. Stutchbury and Morton 2001).

In addition to home ranges, population densities of each bird species studied here are estimated in Terborgh et al. (1990), and I compared my results with that study's (Table 3.14). Population densities of all study species dropped significantly throughout the present study, but estimates here are generally much higher than those calculated in Terborgh et al. (1990). The lowest population densities of

TABLE 3.14. A comparison of population estimates and home-range sizes for five species of obligate ant-following birds at Cocha Cashu Biological Station, Peru. The 1990 estimates are from Terborgh et al. (1990); the 1998–2001 estimates are from the present study.

	1990			1998–2001		
Species	Number of individuals per 100 ha	Home range (ha)	*n*	Number of individuals 100 per ha	Mean home range (ha)	*n*
D. merula	16	NA	–	16.7–22.1	64.6	20
P. nigromaculata	9	14	3	11.3–18.3	45.9	20
M. fortis	2	>50	1	10.4–11.7	15.4	4
R. melanosticta	4	>25	2	5.4–12.5	50.8	24
G. salvini	1	>25	1	8.8–17.5	27.5	21

three species (*D. merula*, *P. nigromaculata*, and *R. melanosticta*) were fairly similar to the densities found in 1982; the other two species (*M. fortis* and *G. salvini*) had population lows here that were 4.3–7.7× higher than the 1982 density estimates. Although population sampling was low in Terborgh et al. (1990), the large discrepancies between some results may represent natural population fluctuations in the obligate ant-following birds. Evidence is presented in Chapter 6 that overall population fluctuations are attributable to both low juvenile recruitment and variable adult survival. More years of continuous study will determine whether the trend of overall population decrease observed from 1998 through 2001 is part of a larger cyclical population fluctuation.

SEASONALITY AND MOVEMENT PATTERNS

Individual movement patterns, and thus average home-range sizes, increased markedly for all bird species sampled in 1999. *Eciton burchelli* density was not abnormally low in that year, and total population density of obligate ant-followers was intermediate between 1998 and 2000–2001 in the study area (Table 3.3). Anomalous rainfall patterns from Cocha Cashu correspond with the expansion in home-range size across bird species. The rainy season began one month earlier in 1999 than in 1998 or 2000, and 1204.5 mm of rain fell from September to December, as compared with 741 mm in 1998 and 868 mm in 2000 (Fig. 3.4). The period of maximal rainfall was in November and December in 1999–2000, two months earlier than in 1998–1999 and 2000–2001, when maximal ranfall was in January and March. The study plot lies entirely within floodplain forest, which, with inundation and rising river levels, becomes a mosaic of wet and dry patches in the rainy months (see Terborgh 1983).

In the 1999 early rainy season, dry land was fragmented into temporary islands within a matrix of rainwater over much of the study plot. Standing water had an observable effect on army ant colonies, which had to "hopscotch" from dry land-mass to dry land-mass. Although *E. burchelli* density in 1999 was similar to that of other years (Table 3.1), much foraging during the period occurred above ground, in shrubs, trees, and vine tangles—unsuitable for avian foragers. *Labidus praedator* is dependent on dry land, because most of its foraging is done underground or on the surface of the ground, and colony nests are located underground. For those reasons, it was impossible to obtain an accurate measure of swarm density for *L. praedator* in 1999. Although *L. praedator* colony density may have been similar to that of other years, it is probable that colonies were heterogeneously distributed within the study plot as a result of inundation.

The decrease in foraging opportunities available to birds at *E. burchelli* swarms in 1999 was evident from the increased rate of agonistic encounters at *E. burchelli* swarms in that year (Tables 3.9 and 3.10). That increase is notable because bird populations had decreased from 1998, whereas army ant colony density was slightly higher in 1999 than in 1998, which suggests that less competition should have occurred (Tables 3.1 and 3.3). Fewer foraging opportunities may have forced obligate ant-followers to maintain larger home ranges so as to encompass enough ground-foraging swarms.

In the 2000–2001 field season, mean home-range sizes of all bird species were similar to those seen in 1998, before the study plot was inundated. More years of study in both the dry

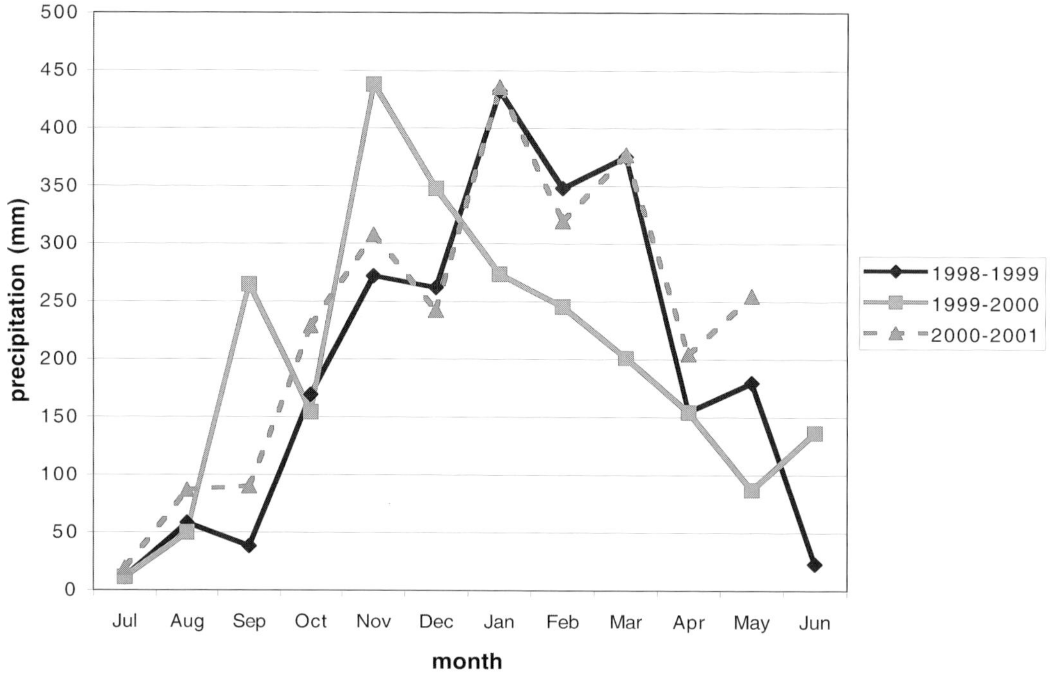

FIG. 3.4. Monthly rainfall at the Cocha Cashu Biological Station, Manu National Park, Peru, from 1998 to 2001. Graph organization follows the cyclic rainy and dry seasons. Precipitation patterns were similar for the rainy seasons of 1998–1999 and 2000–2001. Approximately 400 mm more rain fell from September to December of 1999 than in 1998 or 2000, though total rainfall in each rainy season was similar (2,322, 2,363, and 2,607 mm).

and wet seasons would certainly help to clarify the connections between rainfall and movement patterns of obligate ant-following birds.

CONCLUSION

In summary, species coexistence among five obligate ant-followers is at least partially explained by their differential use of the two army ant species, *E. burchelli* and *L. praedator* (Tables 3.7 and 3.8). Most species minimized home ranges well below the size required to follow multiple *E. burchelli* colonies by relying on unpredictable but abundant foraging opportunities at *L. praedator* swarms (Table 3.6). By using both *L. praedator* and *E. burchelli* swarms, individual birds maximized foraging space per bird (Table 3.13) and decreased agonistic interactions between individuals. Further decreases in potential interspecific competition were gained through one species' specialization on *L. praedator* army ants (Fig. 3.5) and another species' use of white-lipped peccary herds. The three obligate antbirds that used *E. burchelli* in relatively equal proportions are segregated by body mass (Table 3.4), which may allow differential use of space along the width of a swarm front (Fig. 3.3).

CONSERVATION CONCERNS

Although the birds in Manu National Park, Peru, are in no danger from forest fragmentation, obligate ant-followers in other areas of the tropics have been shown to be one of the first groups of birds to go locally extinct once an area has been isolated (Harper 1989, Stouffer and Bierregaard 1995). That seems to be attributable, in part, to their requirements of relatively large home ranges, as compared with birds of similar size, and also to their complete reliance on army ants. *Eciton burchelli*, because of its narrow tolerance for changes in humidity, requires forest with canopy cover (Schneirla 1971). If the ants die out, the birds either die out as well or, as demonstrated in the Biological Dynamics of Forest Fragmentation project in Brazil, abandon fragments of ≤100 ha for mature forest (Harper

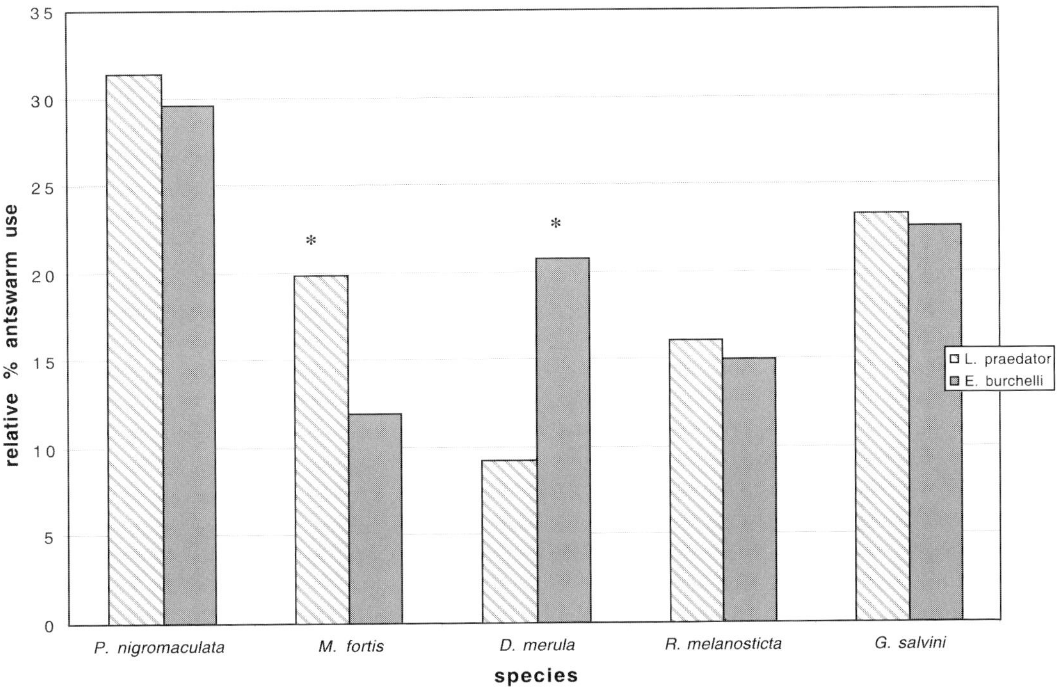

FIG. 3.5. Of five obligate ant-following bird species in Amazonian Peru, *M. fortis* and *D. merula* utilize the army ants *E. burchelli* and *L. praedator* in significantly different proportions (chi-square tests: $P < 0.001$).

1989, Stouffer and Bierregaard 1995, Bierregaard et al. 2002).

In Panama, flooding of the Chagres River to form the Panama Canal and Gatun Lake has led to local extinctions of some obligate ant-followers (Willis 1974, Robinson 1999). The dominant ant-follower in Panama, *Pha. mcleannani*, is one of many avian species that have gone locally extinct on the 1,600-ha island of Barro Colorado (BCI) since it was formed in 1913. The smaller obligate species *G. leucaspis* is now found at a density of ~2.5 individuals per 100 ha on BCI, compared with 12 individuals per 100 ha on the nearby mainland, and is predicted to go locally extinct on BCI in the near future (Robinson et al. 2000a, W. D. Robinson pers. comm.).

The island has had a stable population of *E. burchelli* colonies for >50 years, estimated at 3.2 colonies per 100 ha (Willis 1967, Franks 1982b). Willis (1967) estimated the average *L. praedator* swarm density over 14 months as 8.5 active swarms per 100 ha (no recent estimates are available for BCI). However, *L. praedator* swarm density on BCI fluctuates dramatically from wet to dry season (Willis 1967), which suggests that reliance on *L. praedator* swarms is an unlikely strategy on BCI. Equation 3.4 predicts that *Pha. mcleannani* and *G. leucaspis* on BCI hold larger home ranges than obligate ant-followers at Cocha Cashu, because of lower *E. burchelli* colony density on BCI; however, the island is still predicted to hold >750 individual *Pha. mcleannani*, assuming that conditions approximate Panama's mainland (using the estimate of 48 individuals per 100 ha on the "Limbo plot"; Robinson et al. 2000a, W. D. Robinson pers. comm.). However, that assumption is not valid. Experimental tests have determined that nest predation rates on BCI are significantly higher than relative rates on the mainland, possibly because of mesopredator release (Loiselle and Hoppes 1985, Sieving 1992). Furthermore, Sieving (1992) found that BCI-extinct birds that are terrestrial insectivores have relatively predator-safe nests on the mainland; as a result, she argued, those species may have evolved lower capacity to renest; unable to adapt to the high nest-predation pressures on BCI (brought about by isolation and mesopredator release), they went locally exinct. Additionally, if high annual population fluctuations found in Cocha Cashu birds may be generalized, they provide another potential reason for the disappearance of *Pha. mcleannani* from BCI. As an island population, *Pha. mcleannani* on BCI lacks a source population from which to add more individuals after "bad" years (Willis 1974, Leigh 1982, Robinson 1999).

Insights from studies of bird communities in Panama and Brazil highlight the sensitivity of obligate ant-following birds to fragmentation, isolation, and nest predation. A working understanding of the ecological needs of this specialized, vulnerable guild of birds may facilitate conservation plans that will prevent the obligate ant-followers from going locally extinct as larger areas of lowland rainforest are continually altered and fragmented by humans.

4. SURVIVAL RATES AND POPULATION DYNAMICS IN OBLIGATE ARMY ANT FOLLOWERS

LOWLAND TROPICAL RAINFORESTS have historically been characterized as climatically benign environments (MacArthur 1972). In comparison with the wide fluctuations in temperature and day-length associated with temperate areas, tropical lowlands display seemingly insignificant seasonal changes in those factors. That stability is the backbone of many early "equilibrium" theories put forth to explain the high species diversity in the tropics (Dobzhansky 1950, Fischer 1960, Connell and Orias 1964, Pianka 1966, MacArthur 1969). In theory, climatic stability promotes year-round resource stability, which leads to increased specialization and thus to increased coexistence of ecologically similar species (Orians 1969, Leigh 1982). A stable resource-base leads to population stability in consumers, whose populations are theoretically most affected by density-dependent factors and held at or near a stable carrying capacity (Cody 1966, MacArthur 1972, Newton 1998).

However, lowland tropical systems display distinct fluctuations in some abiotic parameters; in particular, ecologists have documented the importance of seasonal rainfall variation. The transition from rainy to dry season coincides with significant drops in leaf-litter and canopy arthropod abundance (Williams 1941, Janzen 1973, Willis 1976, Wolda 1978, Levings and Windsor 1982, Smythe 1974, Pearson and Derr 1986), leaf loss in deciduous canopy trees (Leigh and Windsor 1982), and a period of famine in many mammal and bird species (Leigh and Windsor 1982, Leigh 1999). Large fluctuations in the tropical resource-base seem at odds with the theoretical underpinnings of tropical stability; some studies question whether the "stable" tropics are as stable as ecologists once proposed. Karr and his colleagues (Karr 1971, Karr et al. 1982, Karr and Freemark 1983) described how bird populations can fluctuate seasonally with changing resource levels; they found that individuals continually reselect microhabitat in response to gradients in moisture and vegetation. Karr and Freemark (1983) asserted that tropical bird assemblages are highly dynamic in space and time, leading to nonequilibrium but still resulting in relatively predictable communities. Issues of scale must be considered; Karr's work was carried out on a 2-ha study plot. Terborgh et al. (1990) challenged whether the scale of Karr's work was sufficient to adequately sample the bird community, citing primarily the fact that the mean home-range size for tropical birds is larger than Karr's study plot.

In addition to seasonal fluctuations, the lowland tropics undergo periodic climatic disturbances, most commonly through cyclic El Niño Southern Oscillation (ENSO) processes (Condit et al. 1996, Wright et al. 1999, Holmgren et al. 2001). Janzen (1967) discussed tropical stability as promoting narrow ecological tolerances to gradients in temperature and rainfall in tropical species as compared with temperate species. Although his arguments were made to explain patterns in tropical species' ranges, they have implications for population dynamics. In tropical dry forest, Faaborg (1982) found that population levels of some resident bird species dropped significantly in response to drought. Stiles (1992) reported how an unusual drought affected survivorship, breeding, and body condition in a rainforest population of *Phaethornis superciliosus* hummingbirds. In that study, recovery of age-structure and population took 3–4 years. Effects of ENSO in the southwestern Amazon basin are small, and southeastern Peru, where the present study took place, is minimally affected because it is located just south of the transition node between the wetter-than-normal conditions of northern South America and the drier-than-normal conditions of Bolivia and southwestern Brazil (Marengo 1992, Pezzi and Cavalcanti 2001, Coehlo et al. 2002, Silman et al. 2003). In the southwestern Amazon, cold air-masses from far-southern-latitude storms periodically move north along the Andean chain and descend into the Amazon basin in the dry season, pushing interior forest temperatures below 15°C for up to five days (Terborgh 1983). In the rainy season, destructive wind- and rainstorms move through lowland tropical forests unpredictably, creating new gaps the size of football fields (Vandermeer et al. 2000). During that season, which coincides with the nesting seasons of most insectivorous birds, rapid flooding can create a mosaic of dry and wet patches within floodplain forest.

In some species, unpredictable fluctuations in abiotic factors may influence juvenile and adult mortality, reproduction, and population and community dynamics (Stouffer and Bierregaard 1993, 1996).

In the southwestern Amazon basin, obligate ant-following birds are exposed to seasonal rainfall patterns and to irregular cold fronts and floodplain inundation. All those factors may affect density and availability of army ants, the ant-followers' foraging resource. In the previous chapter, I asked how the five species of obligate ant-followers at Cocha Cashu differentially used the resources offered by two species of army ants (*L. praedator* and *E. burchelli*). Two of the five bird species (*D. merula* and *M. fortis*) preferentially foraged with one or the other of the two ant species; those preferences showed up in the specific behavioral ecology of each bird species. *Myrmeciza fortis* pairs held small conspecifically defended territories, which allowed them exclusive access to the *L. praedator* swarms in their territory. On the other hand, the woodcreeper *D. merula* preferred *E. burchelli* swarms. Ecologically, assessment of multiple antswarms allowed *D. merula* individuals to maintain home ranges that averaged 5× larger than the similarly sized *M. fortis*. Through linear regression analyses, I found that obligate ant-following species at Cocha Cashu used different amounts of space while foraging along an antswarm front—which facilitates inclusion at swarms of smaller, more subordinate species that can "fit" into and forage effectively in the spaces around their larger competitors. Finally, I found that *D. merula* supplemented its foraging with army ants by following herds of white-lipped peccaries.

All those ecological and behavioral differences enhance the coexistence of the five species of obligates at Cocha Cashu, but there are still unanswered questions regarding species coexistence. Most Neotropical bird species that have been intensively studied have stable populations, as well as stable territorial boundaries, which often remain unchanged for years (Willis 1974; Munn and Terborgh 1979; Munn 1985; Greenberg and Gradwohl 1986, 1997; Roper 1996; Jullien and Thiollay 1998; Robinson 2000; Styrsky 2003), and territoriality has been hypothesized as a driving proximate force in regulating populations (Greenberg and Gradwohl 1986). In contrast to generalist territorial insectivores, most obligate ant-following species hold much larger, more amorphous home ranges that are not conspecifically exclusive (see Chapter 3; also Willis 1967, 1973). If territoriality is an important proximate factor that keeps local bird populations stable, obligate ant-followers should show less rigidity in their population dynamics. Ant-following birds aggregate at food resources, and population fluctuations have the potential to affect rates of agonistic encounters between dominant and subordinate species. Ecologically similar species may be affected by changes in population levels of their competitors, and periodic fluctuations in those levels may contribute to maintaining high diversity within a guild.

Here, I explore the population dynamics of a guild of five obligate ant-following bird species in southeastern Peru over a five-year period. My data on adult annual survival rates, population densities, food-resource densities, and rainfall enabled me to assess the effects of army ant density (a biotic factor) and of precipitation (an abiotic factor) on bird population parameters. I investigated four questions: (1) What are the estimated adult annual survival rates and recruitment rates for the five species of obligate ant-followers? (2) Do annual rainfall patterns or antswarm density correlate with changes in avian population parameters? (3) Does a strictly territorial ant-following species display population dynamics that differ from those of ant-following species with less rigid (and non-exclusive) home ranges? (4) Can fluctuations in population dynamics of individual species assist in maintaining coexistence of the five obligate ant-followers?

Methods

Survival Estimates

Intensive capture-and-resight data collected over five years were used to estimate annual adult survival rates. Data were analyzed in the program MARK (White and Burnham 1999), following the approach outlined by Lebreton et al. (1992). Cormack-Jolly-Seber (CJS) constant and time-dependent survival and recapture models were fitted for all species. In addition, I include models built *a priori* to test specific hypotheses about survival and recapture rates for particular species. *A priori* models were based on the following observations: (1) the study plot was not identical in 1999 and 2000, and thus may have altered recapture rates;

(2) adult *R. melanosticta* sightings were much lower in 2000; and (3) capture-and-resight effort was lower in 2002. All survival intervals are denoted by the year in which the interval ends (e.g. 2000 denotes the interval 1999–2000). The *a priori* hypotheses were as follows:

(1) Survival rate (ϕ) is considered constant, and recapture probability (*P*) is different in 2002 than in other years. This model accounts for the shorter duration of the 2002 field season, which may have lowered bird resighting rates. Model 1 is ϕ(.)p(2002).

(2) Survival rate for the interval 1999–2000 is different from that of other years, and recapture rate is constant. This model formally tests whether survival rate was different in the 1999–2000 interval as compared with other intervals, and is based on the *a priori* observed drop in number of *R. melanosticta* individuals between the 1999 and 2000 seasons. Model 2 is ϕ(2000)p(.).

(3) Survival rate for the interval 1999–2000 is different from that of other years, and recapture rate is time dependent. This model is similar to model 2, but predicts a different recapture rate for each year interval. Model 3 is ϕ(2000)p(t).

(4) Survival rate is constant, and recapture rate in 2000 is different from that of other years. This model accounts for the fact that the study plot was not identical in 1999 and 2000, and thus may have altered recapture rates. Model 4 is ϕ(.)p(2000).

(5) Survival rate is constant. Recapture rate is the same between 1998 and 1999, and differs in both 2000 and 2002. This model formally tests whether the observed drop in number of *R. melanosticta* individuals between 1999 and 2000 is attributable to differences in recapture rate, with the assumption that 1998 and 1999 had similar recapture rates. Recapture rate is modeled separately for 2002 to account for the shorter duration of the field season. Model 5 is ϕ(.)p(1998 = 1999, 2000, 2002).

(6) Survival rate is constant. Recapture rate is the same between 1998 and 1999, and the same between 2000 and 2002. This model accounts for the fact that the percentage of *G. salvini* color-banded individuals was much higher in 2000–2001 and 2002 than in 1998 and 1999. Model 6 is ϕ(.)p(1998 = 1999, 2000 = 2002).

Models were selected for fit on the basis of number of parameters and deviance; the most parsimonious model has the lowest value for Akaike's Information Criterion (Burnham and Anderson 2002). Models that did not support *a priori* hypotheses were not selected as candidate models to describe the data. In accordance with White (2002), I used the program RELEASE to test overdispersion in the data, rather than relying on bootstrap goodness-of-fit tests. Data for four of the five species displayed no overdispersion. A fifth species, *M. fortis*, holds small territories; thus, sample sizes of banded birds of that species for survival estimates are low (n = 15). To avoid overfitting the data, only the four basic CJS models were run on the *M. fortis* data set, and results are considered preliminary.

Annual adult population densities with 95% confidence intervals were calculated using the Bowden model-estimation technique in the program NOREMARK (White 1996; see Chapter 3 for more details). Recruitment rate for each species was estimated for the 1999 and 2000 breeding seasons. I multiplied adult survival rate by estimated adult population density in year t (calculated with Bowden's model estimator on field data) to estimate the adult population, or "survivors," still alive the following year (year $t + 1$). That estimate was then compared with the estimated adult population as gathered from field data for that year (year $t + 1$), calculated with Bowden's model estimator. The replacement rate is the Bowden's estimated population minus the estimated survivors. A replacement rate of zero equals no recruitment; a negative number signifies no recruitment plus some adult mortality above that predicted by the estimated adult survival rate; and a positive number signifies recruitment into the population, either by young of the previous year or by adult floaters from outside areas.

Results

Survival and Recruitment Rates

The best models for four of the five study species had constant annual adult survival over the five-year period. Survival was higher in larger-bodied species (Table 4.1). *Dendrocincla merula* survival was highest, at ϕ = 0.80; *M. fortis* survival was estimated at ϕ = 0.72, *P. nigromaculata* at ϕ = 0.63, and *G. salvini* at ϕ = 0.59. *Rhegmatorhina melanosticta* supported a model with a constant adult survival rate in the first, second, and fourth yearly intervals, at ϕ = 0.62, but survival dropped to ϕ = 0.36 in the interval 1999–2000.

Estimated recruitment rates varied widely by year and species (Table 4.2 and Fig. 4.1). From 1998 to 1999, the total adult population declined by 23% (see Table 3.3), and three species (*P. nigromaculata*, *R. melanosticta*, and *D. merula*) showed no recruitment. Based on confidence intervals, population estimates for two of those species—*P. nigromaculata* and *R. melanosticta*—showed significant declines in the same period, whereas the *D. merula* population decline was not as clear. In the same interval, *M. fortis* showed positive recruitment and *G. salvini* showed high recruitment (6.4 individuals per 100 ha; Fig. 4.1).

From 1999 to 2000, the total population of obligate ant-followers decreased again, from

TABLE 4.1. Model selection results from the program MARK for apparent adult survival (ϕ) and recapture rates (p) for five ant-following bird species in Amazonian Peru between 1997 and 2002; CI = 95% confidence interval.

Dendrocincla merula

Model	AIC_c	ΔAIC_c	AIC_c weight	#Par	Deviance
{ϕ(.) p(2000,2002)}	193.005	0.00	0.319	4	18.613
{ϕ(.) p(t)}	193.070	0.06	0.308	5	16.525
{ϕ(2000) p(t)}	194.536	1.53	0.148	6	15.804
{ϕ(t) p(t)}	195.384	2.38	0.097	7	14.431
{ϕ(t) p(.)}	195.733	2.73	0.081	5	19.187
{ϕ(.) p(98=99,00=02)}	198.093	5.09	0.025	3	25.823
{ϕ(.) p(2002)}	198.447	5.44	0.021	3	26.177
{ϕ(.) p(.)}	209.925	16.92	0.000	2	39.745
{ϕ(2000) p(.)}	211.266	18.26	0.000	3	38.996
{ϕ(.) p(2000)}	211.319	18.31	0.000	3	39.049

Real function parameters of {ϕ(.)p(2000,2002)}

Parameter	Estimate	SE	Lower CI	Upper CI
1:ϕ	0.799	0.050	0.682	0.880
2:p	0.974	0.026	0.837	0.996
3:p	0.697	0.105	0.465	0.859
4:p	0.354	0.098	0.191	0.560

Phlegopsis nigromaculata

Model	AIC_c	ΔAIC_c	AIC_c weight	#Par	Deviance
{ϕ(.) p(.)}	157.271	0.00	0.250	2	15.877
{ϕ(.) p(98=99,00=02)}	157.593	0.32	0.213	3	14.063
{ϕ(2000) p(.)}	158.307	1.04	0.149	3	14.776
{ϕ(.) p(2002)}	158.742	1.47	0.120	3	15.211
{ϕ(.) p(2000)}	159.003	1.73	0.105	3	5.473
{ϕ(.) p(2000,2002)}	159.502	2.23	0.082	4	3.787
{ϕ(t) p(.)}	161.189	3.92	0.035	5	13.239
{ϕ(.) p(t)}	161.607	4.34	0.029	5	13.656
{ϕ(2000) p(t)}	163.086	5.82	0.014	6	12.848
{ϕ(t) p(t)}	165.380	8.11	0.004	7	12.802

Real function parameters of {ϕ(.) p(.)}

Parameter	Estimate	SE	Lower CI	Upper CI
1:ϕ	0.631	0.056	0.516	0.733
2:p	0.838	0.073	0.644	0.936

Myrmeciza fortis

Model	AIC_c	ΔAIC_c	AIC_c weight	#Par	Deviance
{ϕ(.) p(.)}	49.790	0.00	0.804	2	10.387
{ϕ(.) p(t)}	52.836	3.05	0.175	5	4.955
{ϕ(t) p(.)}	57.723	7.93	0.015	5	9.842
{ϕ(t) p(t)}	59.728	9.94	0.006	7	4.625

Real function parameters of {ϕ(.) p(.) PIM coding}

Parameter	Estimate	SE	Lower CI	Upper CI
1:ϕ	0.723	0.108	0.475	0.883
2:p	0.754	0.143	0.405	0.933

TABLE 4.1. Continued.

Rhegmatorhina melanosticta

Model	AIC_c	Δ AIC_c	AIC_c weight	#Par	Deviance
{ϕ(2000) p(.)}	94.889	0.00	0.491	3	6.070
{ϕ(.) p(.)}	96.696	1.81	0.199	2	10.070
{ϕ(.) p(2002)}	98.889	4.00	0.067	3	10.070
{ϕ(.) p(2000)}	98.889	4.00	0.067	3	10.070
{ϕ(.) p(98=99,00=02)}	98.889	4.00	0.067	3	10.070
{ϕ(t) p(.)}	99.124	4.23	0.059	5	5.702
{ϕ(.) p(2000,2002)}	101.153	6.26	0.021	4	10.070
{ϕ(2000) p(t}	101.548	6.66	0.018	6	5.710
{ϕ(.) p(t)}	103.492	8.60	0.007	5	10.070
{ϕ(t) p(t)}	104.038	9.15	0.005	7	5.702

Real function parameters of {ϕ(2000)p(.)}

Parameter	Estimate	SE	Lower CI	Upper CI
1:ϕ	0.622	0.072	0.474	0.751
2:ϕ	0.364	0.103	0.193	0.577
3:p	1.000	0.000	1.000	1.000

Gymnopithys salvini

Model	AIC_c	Δ AIC_c	AIC_c weight	#Par	Deviance
{ϕ(.) p(98=99,00=02)}	125.077	0.00	0.301	3	21.206
{ϕ(.) p(.)}	126.921	1.84	0.120	2	25.215
{ϕ(.) p(t)}	127.171	2.09	0.106	5	18.791
{ϕ(.) p(2000,2002)}	127.300	2.22	0.099	4	21.206
{ϕ(t) p(t)}	127.353	2.28	0.097	7	14.207
{ϕ(.) p(2002)}	127.608	2.53	0.085	3	23.737
{ϕ(.) p(2000)}	127.686	2.61	0.082	3	23.815
{ϕ(2000) p(.)}	129.027	3.95	0.042	3	25.156
{ϕ(t) p(.)}	129.412	4.34	0.035	5	21.032
{ϕ(2000) p(t)}	129.451	4.37	0.034	6	18.722

Real function parameters of {ϕ(.)p(98=99,00=02)}

Parameter	Estimate	SE	Lower CI	Upper CI
1:ϕ	0.586	0.055	0.475	0.689
2:p	0.805	0.112	0.503	0.944
3:p	1.000	0.000	1.000	1.000

an estimated 67 adults per 100 ha, to 46—a decline of 33% (see Table 3.3). However, patterns within species differed from those observed in 1998–1999. In contrast to its reproductive success in 1998–1999, *G. salvini* was the only species to show negative recruitment in the 1999–2000 interval. That estimate suggests little-to-no juvenile recruitment, in addition to higher adult mortality than predicted by normal adult survival rates for the species (Table 4.2). Confidence intervals for *G. salvini* support a decline in adult population density over that period (Fig. 4.2). *Rhegmatorhina melanosticta* is the only species that shows a drop in population in both the 1998–1999 and 1999–2000 intervals. In 1999–2000, the drop was largely attributable to the nearly 50% decline in adult survival; however, some successful recruitment occurred, which kept the population from declining even further (Table 4.2). The two other species that showed negative recruitment in 1998–1999, *P. nigromaculata* and *D. merula*, both displayed positive recruitment in 1999–2000. In all of the above cases, except for *R. melanosticta* in 1999–2000, there is no change in estimates of adult survival; population declines in the other species appear to be attributable to failure of

TABLE 4.2. Recruitment estimates over two years for five species of obligate ant-following birds in Amazonian Peru. Multiplying the estimated adult survival rate by the estimated population per 100 ha provides an estimate of the adult population still alive the following year. That estimate (e.g. "1999 estimated survivors") is then compared with the actual estimated population per 100 ha for that year (using Bowden population models). The replacement rate is the Bowden estimated population (e.g. "1999 estimated population") minus the estimated survivors. A replacement rate of zero equals no recruitment, a negative number signifies no recruitment plus some adult mortality, and a positive number signifies recruitment into the population, either by young of the previous year or by adult floaters.

			1998–1999		
Species	Survival rate	1998 estimated population	1999 estimated survivors	1999 estimated population	1999 replacement rate
D. merula	0.80	19.8	15.8	15.2	–0.64
P. nigromaculata	0.63	21.0	13.2	13.0	–0.23
M. fortis	0.72	9.3	6.7	10.1	3.40
R. melanosticta	0.62	16.7	10.4	10.1	–0.25
G. salvini	0.59	19.8	11.7	18.1	6.42
			1999–2000		
Species	Survival rate	1999 estimated population	2000 estimated survivors	2000 estimated population	2000 replacement rate
D. merula	0.80	15.2	12.2	14.0	1.84
P. nigromaculata	0.63	13.0	8.2	10.7	2.51
M. fortis	0.72	10.1	7.3	8.5	1.23
R. melanosticta	0.36	10.1	3.6	4.8	1.16
G. salvini	0.59	18.1	10.7	7.7	–2.98

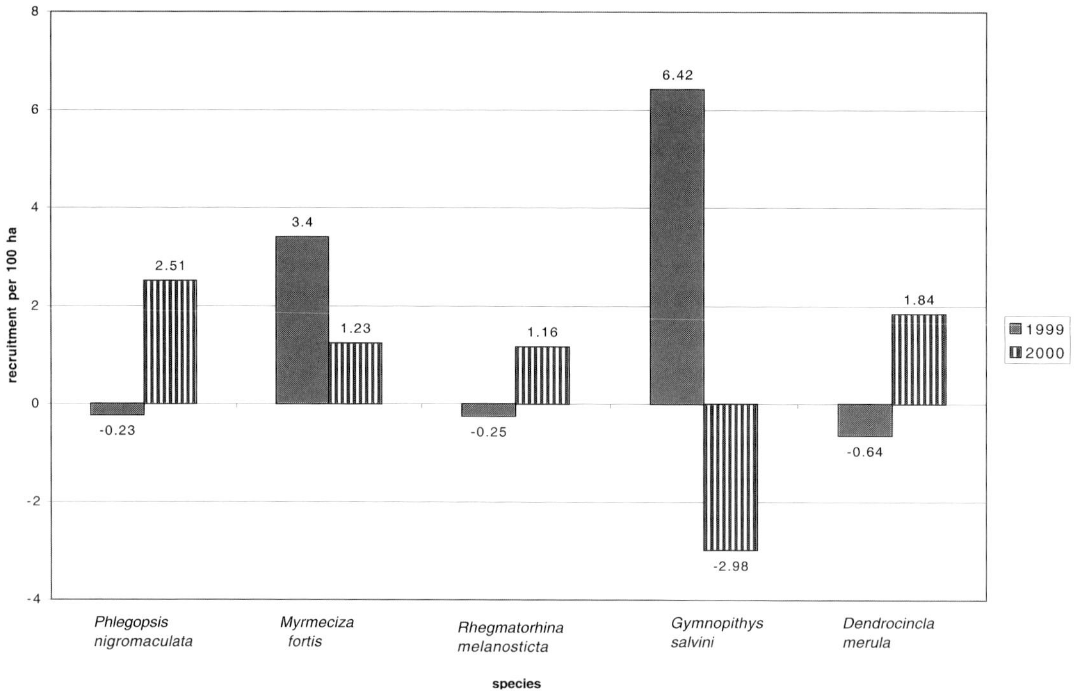

FIG. 4.1. Estimated recruitment rate per 100 ha per year for five obligate ant-following bird species in Amazonian Peru. Recruitment rate is adjusted for adult annual survival; a rate of 0 signifies no juvenile recruitment, a negative rate signifies no juvenile recruitment and adult mortality above the average survival rate, and a rate >0 signifies positive recruitment.

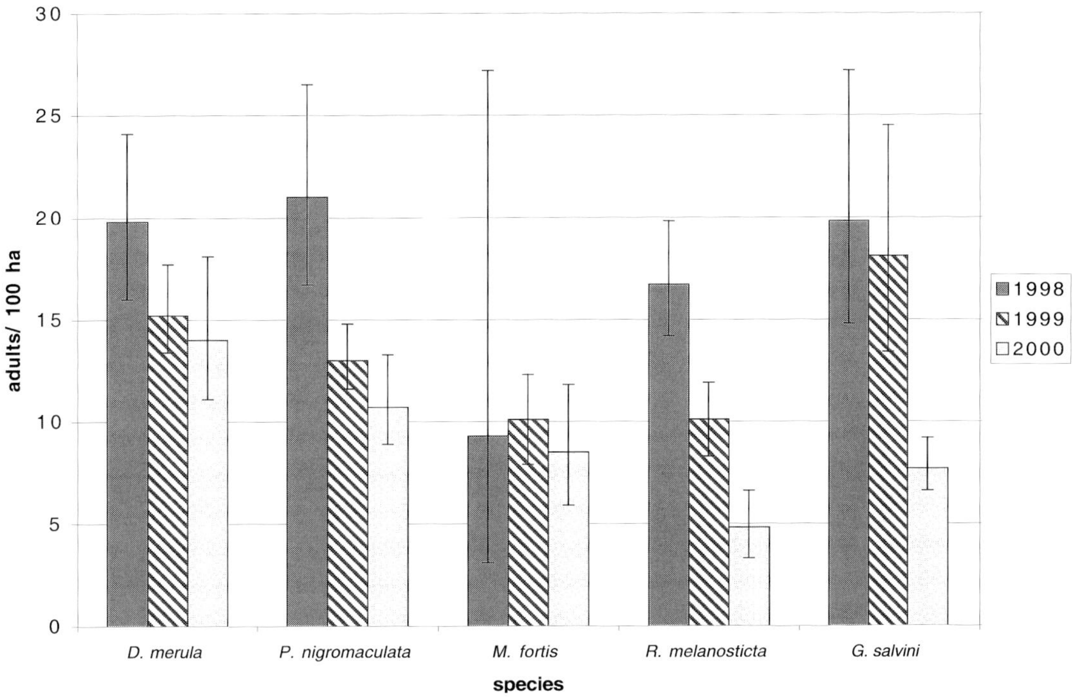

FIG. 4.2. Estimated adults per 100 ha for five species of obligate ant-following birds in Amazonian Peru, over three years. Error bars are 95% confidence intervals, using Bowden's model estimation.

juveniles from the previous breeding season to recruit into the population.

POPULATION DYNAMICS AND TERRITORIALITY

If territoriality is a stabilizing proximate factor for populations of tropical birds, species with less-rigid territorial systems can be hypothesized to show more variation in population dynamics. That can be tested among the four phylogenetically similar obligate antbird species, three of which hold amorphous, non-exclusive home ranges (*P. nigromaculata*, *R. melanosticta*, and *G. salvini*) and one of which is conspecifically territorial (*M. fortis*) (Fig. 4.3). Based on confidence intervals, each of the three nonterritorial antbirds showed significant population declines (Fig. 4.2). Population density of *M. fortis*, on the other hand, remained stable over three years. Because of the small sample sizes, those results should be viewed as preliminary; however, they provide support for the hypothesis that territoriality is a driving proximate force in stabilization of population dynamics in tropical birds (Greenberg and Gradwohl 1986, Newton 1998).

DISCUSSION

Total population density of the five study species declined by ~25% each year, from an estimated high of 87 adults per 100 ha in 1998 to a low of 45 adults per 100 ha in the 2000–2001 field season. Based on confidence intervals, *M. fortis* and *D. merula* adult population levels remained stable throughout the study, *P. nigromaculata* and *G. salvini* adult populations fell in one of the two yearly intervals, and *R. melanosticta* adult populations fell in both intervals (Fig. 4.2). Adult survival estimates for all species were stable throughout the study period, with one exception; only *R. melanosticta*'s drop in population could be attributed to adult mortality, and only for the interval 1999–2000 (Table 4.1). All other drops in population density are attributable to a lack of recruitment of young from the previous breeding season (Table 4.2 and Fig. 4.2). Below, I discuss the implications of low recruitment for obligate ant-followers.

All data from recruitment presented here were inferred from adult survival and population estimates; actual juvenile recruitment was

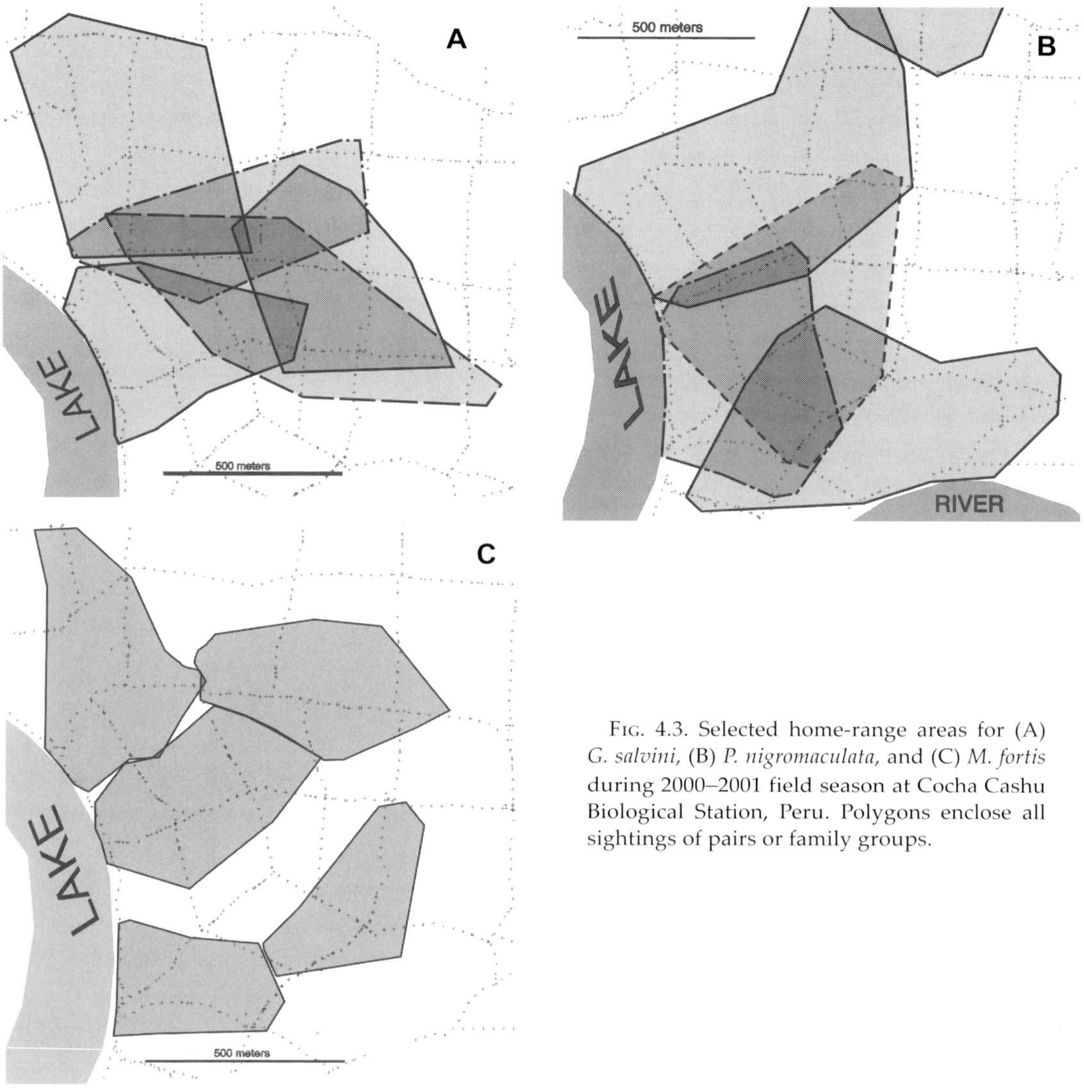

FIG. 4.3. Selected home-range areas for (A) *G. salvini*, (B) *P. nigromaculata*, and (C) *M. fortis* during 2000–2001 field season at Cocha Cashu Biological Station, Peru. Polygons enclose all sightings of pairs or family groups.

not measured. In the few instances where juveniles from one year were resighted in following years with mates (S. K. Willson unpubl. data for *P. nigromaculata* and *R. melanosticta*), individuals were located within a few hundred meters of where they were banded as nestlings. That limited evidence suggests that juvenile dispersal is confined to short distances in some juvenile ant-followers. However, it is also possible that most juveniles dispersed long distances and were not observed again because they settled off the study plot. A study of dispersal using radiotelemetry would provide invaluable data toward answering the question more directly.

To gain a better understanding of potential long-term implications of population fluctuations in these birds, it is instructive to examine rates of recruitment of juveniles in what would be considered "good" years. In the period 1998–1999, the smallest and most subordinate ant-follower, *G. salvini*, showed the highest recruitment and no significant change in adult survival or population density (Table 4.2 and Fig. 4.1). At a level of 19.8 adults, or 10 breeding pairs per 100 ha, during the breeding season of 1998–1999, only 6.4 new recruits entered the population. Those 6.4 adults were presumably birds that fledged during the 1998–1999 breeding season—which signifies that during a "good" year, in which *G. salvini* had the highest recruitment of any obligate ant-follower, each pair contributed an average of 0.6 successful young out of a clutch size of two eggs (Skutch 1996). Between 1999 and 2000, no new fledglings are estimated to have recruited, and the adult population declined significantly.

In the period 1999–2000, highest recruitment was seen in *P. nigromaculata* (Fig. 4.1). Out of an estimated population of 13 adults per 100 ha, 2.5 new juveniles were added to the population. If we assume that, for this cooperative breeder, each breeding group consists of three adults, then 4.3 breeding groups per 100 ha each contributed an average of 0.6 successful young. In the previous period, 1998–1999, no recruitment was seen and a significant adult population decline was evident. Those results are important, because they provide evidence that in obligate ant-followers, juvenile recruitment may be consistently low, and often zero. If recruitment remains low or negative for consecutive years in a closed population, population densities will decrease as a result of normal mortality. In the five ecologically similar study species, population densities fluctuated for almost all species, and total population density within the guild dropped consecutively over three years. Those results suggest that population processes were not local, or recruitment should have been buffered from outside the plot. If we hypothesize that density-dependent effects are driving the system, we can predict that resource abundance or availability should have decreased over the course of the study (MacArthur 1972, Connell 1978, Newton 1998).

How do those population changes correlate with the biotic and abiotic influences of army ant colony density and rainfall? Given that *E. burchelli* colony density did not vary significantly from year to year, army ant colony density in itself likely had little influence on avian population dynamics. Precipitation peaks, on the other hand, were shifted forward in one of the three rainy seasons (September 1999–April 2000) and correlate temporally with the drop in adult survival of *R. melanosticta* (Fig. 3.4 and Table 4.1). During each rainy season, floodplain forest at Cocha Cashu becomes patchily inundated, creating heterogeneous "islands" of dry ground amid wet depression areas that fill with approximately 0.1–1.0 m of standing water (S. K. Willson pers. obs.). When floodplain forest becomes inundated, *E. burchelli* army ants continue to use the area but forage more often above the terrestrial leaf litter, in vegetation and trees. Army ant density thus remains the same, but availability to ant-following birds changes considerably. Aboveground swarms flush fewer insects for birds, because foraging shifts from fan-shaped raids over leaf litter to column-raids up trees (Schneirla 1971). Although total rainfall was similar in the three rainy seasons (September–April in 1998–1999, 1999–2000, and 2000–2001), the shift forward in onset of precipitation in 1999, coupled with the sudden drop in precipitation late in the rainy season, may have disrupted normal foraging patterns in ant-following birds in that year. Concurrently with those changes in precipitation in 1999, all bird species for which we have sufficient data (four of the five) displayed large increases in mean home-range area (see Fig. 3.2). Across species, there was a significant increase in home-range size for 1999 as compared with 1998 and 2000 (ANOVA: $F = 12.56$, df = 2 and 86, $P < 0.0001$). It is possible that some birds moved from the floodplain study-plot to *terra firme* forest during the 1999 wet season. However, of 147 individuals of the five species that were banded by 1998, only one (a *P. nigromaculata* female) was resighted in later years but not in 1999. If any birds left the plot in 1999, they did not return in later years. In "ordinary" years, when inundation is most prevalent in the late rainy season, patchy flooding and antswarm availability may be a normal factor of mortality for juvenile ant-followers that are just learning how to effectively find swarms and forage on their own. During those years, inundation may have less effect on adult ant-followers, which generally complete nesting by the time inundation peaks and thus have lowered energetic demands, coupled with a clear spatial map of nearby army ant colonies (which young birds lack). Of course, dry-season population crises could also have occurred in months when I was not present.

Shifts in precipitation levels, lower availability of *E. burchelli* swarms, and subsequent increase in avian home-range size may account for the sharp drop in adult survival seen in *R. melanosticta* from 1999 to 2000. That subordinate species received the most interspecific aggression from dominant obligate ant-followers at swarms (Table 3.4). Fully 45% of *P. nigromaculata* displacements and 53% of *M. fortis* displacements were directed at *R. melanosticta*. Additionally, *R.*

melanosticta individuals were highly aggressive intraspecifically (52% of total displacements). Competition for foraging positions at swarms with a high prey-base for birds was extremely intense during November–December 1999. Based on rates of displacements per meter per obligate antbird, there was significantly higher interference competition along *E. burchelli* swarm fronts in November–December 1999 as compared with September 1999 ($t = 4.27$, df = 165, $P = 0.001$). If *R. melanosticta* individuals were repeatedly chased away from available swarms by dominant competitors during that period of intense interference competition, some individuals may have starved to death from lack of feeding opportunities, even after increasing their home-range size by a magnitude of 2–3× (see Table 3.5). As home-range size increases, birds may become less efficient at finding ants and face more competition from conspecifics that are likewise increasing their home ranges.

Regular seasonal inundation may be one factor affecting the observed lack of recruitment of young birds, even in what appear to be "good" years (Table 4.2). Population declines resulting from lack of recruitment were evident in three of the five species over three years, and suggested in a fourth species. In long-lived sedentary tropical birds, recruitment will vary from year to year, depending on biotic and abiotic factors. However, recruitment is generally believed to be low in tropical birds (Fogden 1972, Skutch 1985, Morton and Stutchbury 2000, Robinson et al. 2000b). Rainfall patterns, opportunities to find prey at terrestrially foraging *E. burchelli* and *L. praedator* swarms, and other factors related to nesting success and nest predation may affect each species synergistically. The variable environment of Cocha Cashu's floodplain forest may contribute to each species' having particular behavioral and ecological adaptations that allow it be more successful than others under certain environmental and temporal conditions (MacArthur and Levins 1967, Pianka 1981, Abrams 1983, Ives 1995). An underappreciated method of diversification and coexistence of ecologically similar species may be their manner of tolerating fluctuations in the environment (Janzen 1967). By periodically depressing total population and varying the relative abundance of different obligate ant-following species over time, such factors may contribute to the high number of species able to share the "obligate army ant follower" niche at Cocha Cashu.

The floodplain forest of Cocha Cashu reminds us that although tropical climates are certainly "stable" when compared with the seasonal fluctuations of the temperate zone, population stability may be unattainable in some specialized tropical guilds, and particularly in species that are nonterritorial. Leigh (1982) stated that "presumably, the stabler the environment, the more specialized a species can be and still survive, and the more species can coexist." In the case of obligate ant-followers, the instability of the floodplain environment may be acting to increase species diversity by periodically depressing some species' populations.

5. NESTING AND REPRODUCTION IN A GUILD OF OBLIGATE ARMY ANT FOLLOWERS IN AMAZONIAN PERU

New data on nesting and reproduction of Neotropical antbirds (Thamnophilidae)—and, to a lesser extent, of woodcreepers (Dendrocolaptidae)—are accumulating quickly (del Hoyo et al. 2004). However, there is still much to learn regarding nest structure, incubation, and adult behavior within those families. Many species have never been scientifically studied, and basic reproductive parameters are still unknown. Antbirds and woodcreepers are endemic to the New World tropics and their centers of diversity are located in the Amazon Basin. The 209 antbird species have a diversity of nest types—including open cup, enclosed hanging pouch, open-topped cavity, and oven-shaped nests (Skutch 1996, del Hoyo et al. 2004)—with nest heights ranging from ground to canopy. All 52 species of woodcreepers are assumed to be secondary hole nesters (del Hoyo et al. 2004).

Within the specialized guild of obligate ant-following antbirds and woodcreepers, many species' nests remain undescribed. At the start of the present study in 1997 (Wilkinson and Smith 1997), only one of the five study species (*M. fortis*) had a described nest. *Myrmeciza fortis* nests on the ground in covered dome-nests. Few nests have been found of the many *Myrmeciza* species, but *M. exsul*, of Central America south to Ecuador, builds a concealed cup nest "amid low vegetation atop [a] loose foundation of low plants" (Zimmer and Isler 2003).

Nests of the three other obligate antbird species at Cocha Cashu (*P. nigromaculata*, *R. melanosticta*, and *G. salvini*) are described below. I have published preliminary nest information elsewhere for *R. melanosticta* and *G. salvini* (Willson 2000), but here I present more detailed descriptions of nests, eggs, nestlings, and parental behavior in those species. A nest and eggs of *P. nigromaculata* were described from Colombia (Cadena et al. 2000), but the hatchlings disappeared soon after birth. Here, I describe additional nests for that species, as well as detailed feeding behavior by three attending adults at one nest. We never found a nest of the woodcreeper *D. merula*. However, I have included dates of nesting at Cocha Cashu and information on breeding behavior for that species.

Results

Gymnopithys salvini

Nest.—Three nests were found from 1998 to 2000 (Table 5.1). All nests were located in mature floodplain forest, either on high ground or in low-lying swamp. Nests were in dead palm stumps 41–50 cm in height. All nest stumps were narrow in width (6–9 cm) and contained an open cavity 5–26 cm in depth. Height of the floor cavity varied from ground-level to 45 cm. All cavities were completely open on top, providing an unobstructed view of the nest and no cover from rain. Dead palm-leaf material sparsely lined the bottom of each nest cavity (Fig. 5.1).

Eggs.—One egg was measured, at 16.5 × 23.7 mm. All observed eggs were smooth, with a light-pink matte surface, covered with maroon speckling concentrated at the blunt end. An incubating pair was seen foraging at an antswarm 256 m from its active nest. That swarm observation was made between 0745 and 0930 hours; I have insufficient data to say whether one individual arrived and the other immediately left to resume incubation duties.

Nestlings.—One observed nest was successful, fledging two young. On day 10, young were banded, measured, and weighed (Table 5.2). The female nestling was larger than her male sibling in weight, wing, and tarsus measurements, and bill measurements were similar between the two. The male nestling had one mature botfly larva on the side of his head. Nestlings were sexually dimorphic in color, as are adults. The male was dark gray, with brown wings, while the female was primarily rufous. Parents of nestlings were observed foraging at antswarms ≤216 m from their active nest. One nest was followed from egg stage through fledging; chicks left the nest in the morning 12 days after hatching (Table 5.3).

Fledglings.—Both of the fledglings remained with their parents for at least three months. Immediately after fledging, the chicks hid in dense vegetation behind antswarm fronts and were fed by their parents. On 24 November 2000 (9 days after fledging), one of the fledglings was observed 10 m from an *E. burchelli* swarm

TABLE 5.1. Characters of nests found at Cocha Cashu Biological Station, Peru, for the obligate ant-following antbird species *Rhegmatorhina melanosticta*, *Phlegopsis nigromaculata*, and *Gymnopithys salvini*.

	Nesting substrate	Structure height (cm)	Structure DBH (cm)	Inner cavity diameter (cm)	Inner depth of cavity (cm)	Canopy cover	% concealed from above	% concealed from sides	Nest height above ground (cm)	Habitat	Understory	Bird ID
						R. melanosticta						
Nest 1	*Scheelea* sheath	73 (tree 4.75 m)	42	10.4 × 6.1	10.1	90%	0%	100%	63	High ground, mature	Open	1998Rs-ow
Nest 2a	*Scheelea* sheath	100 (tree 4.25 m)	34	5.6 × 8.0	10.7	95%	0%	100%	89.3	High ground, mature	Open	1999Rs-wp
Nest 2b	[Same nest as 2a; all characters are the same]											
Nest 3	Stump	102	17.2	9	7	95%	90%	100% (S, E, W), 0% (N)	21	High ground, mature	Open	2000Rgp-s
						P. nigromaculata						
Nest 1	Live tree	16–28	17	7.7 × 7.3	10	92%	0%	100%	8	High ground, edge of swamp	Open	2000 Pyb-s
Nest 2	Stump	95	21	7.5 × 4.5	15	65%	50%	100%	51	High ground, gap edge	Closed, brushy	2000Pby-s
Nest 3	Stump	51	10.5	9.7	13	99%	0%	100%	30	High ground	Open	2000 9a
						G. salvini						
Nest 1	Stump	44.5	9.7	6.8 × 7.2	26	85%	0%	100%	0	High ground, *Heliconia*	Herb-aceous	1998 Gs-fg
Nest 2	Stump	41	6.8	5.6	15	90%	0%	100%	23	High ground	Open	2000 Gs-oy
Nest 3	Stump	50	10	9	5	95%	0%	100%	45	Swamp	Open, inundated	2000 Gbw-s

FIG. 5.1. Stump containing nest and two eggs of *G. salvini*.

front. By mid-February, fledglings were generally foraging on their own but still occasionally begged food from their parents. There was no indication that fledglings were divided between parents, with one parent always providing for the same fledgling; Willis (1967) described that type of division in *G. leucaspis* of Panama. Each fledgling *G. salvini* begged and was fed by either parent, and the family remained together at all times while under daily observation. Neither fledgling was seen the following year, although it is possible that they dispersed and were missed during the short field-season of 2002.

Rhegmatorhina melanosticta

Nest.—Four active nests were found from 1998 to 2000 (Table 5.1). One nest site was reused a second time, presumably by the same pair, after an initial successful breeding (see Table 5.3). All nests were located within mature high-ground forest. Three nests were located in vertical cavities left by fallen leaves along the trunk of mature *Scheelea* spp. palm trees (Fig. 5.2); the other nest was in the hollow vertical cavity of a dead sapling. Nest height ranged from 21 to 90 cm above ground. Cavity depth ranged from 7 to 11 cm, and width ranged from 5.5 to 10.5 cm. The nest itself was a shallow bed of dry, shredded palm leaves, completely open at the top.

Eggs.—Three eggs were measured from two separate nests. Eggs averaged approximately 17 × 22 mm and displayed a smooth matte surface, light pink with longitudinal dark-maroon streaks, interspersed with maroon

Table 5.2. Measurements of nestling birds at specified ages for *Rhegmatorhina melanosticta*, *Phlegopsis nigromaculata*, and *Gymnopithys salvini* from Cocha Cashu Biological Station, Peru.

Species	Age (days)	Sex	Weight (g)	Wing (mm)	Tarsus (mm)	Culmen (mm)	Bill width (mm)	Bill depth (mm)	Number of botflies	ID
R. melanosticta	~12	u	23.5	50.0	25.7	5.3	2.9	3.1	0	2000Rr-sy
R. melanosticta	~12	u	21.0	46.0	25.1	6.0	2.9	3.0	0	2000Ry-sy
R. melanosticta	13	f	21.7	55.0	27.2	6.7	3.7	3.9	0	1999Rgw-s
P. nigromaculata	~9	u	33.5	-	25.7	4.7	-	-	7	2000Pr-sb
P. nigromaculata	~9	u	33.0	-	24.1	4.4	2.5	3.0	3	2000Psg-y
P. nigromaculata	~11	u	23.0	46.0	27.0	5.2	3.1	3.1	1	2000Psy-y
P. nigromaculata	~11	u	31.5	53.0	29.1	4.2	3.4	3.2	1	2000Pb-sb
G. salvini	10	m	18.5	43.0	23.3	5.7	3.6	3.3	1	2000Gs-ro
G. salvini	10	f	19.5	48.0	24.2	5.7	3.5	3.0	0	2000Gp-sp

TABLE 5.3. Nesting dates and nest outcomes for three species of obligate ant-following birds in Amazonian Peru.

	Date found, contents	Date complete clutch	Date hatched	Date fledged (or failed)	Total days observed	Total nestling stage (days)	Fledged?	Bird ID	Number of botflies on nestlings
				R. melanosticta					
Nest 1	11 Nov 1998, 1 egg	–	–	13 Nov 2000 (failed)	5	–	No	1998Rs-ow	–
Nest 2a	17 Sept 1999, 2 eggs	–	26 Sept 1999	9 Oct 1999	13	13	Yes (1)	1999Rs-wp	0, 0
Nest 2b	5 Dec 1999, 2 eggs	–	–	–	1	–	–	1999Rs-wp	–
Nest 3	16 Nov 2000, 2 nestlings	–	–	21 Nov 2000	6	–	Yes	2000Rgp-s	0, 0
				G. salvini					
Nest 1	7 Nov 1998, 2 eggs	–	–	20 Nov 1998 (failed)	14	–	No	1998Gs-fg	–
Nest 2	2 Nov 2000, 2 eggs	–	3 Nov 2000	15 Nov 2000	13	12	Yes	2000Gs-oy	1, 0
Nest 3	5 Dec 2000, 2 eggs	–	–	14 Dec 2000 (failed)	8	–	No	2000Gbw-s	–
				P. nigromaculata					
Nest 1	27 Oct 2000, 2 nestlings	–	–	30 Oct 2000	4	–	Yes	2000Pyb-s	1, 1
Nest 2	2 Jan 2001, 2 nestlings	–	1 Jan 2001?	14 Jan 2001	12	13+	Yes	2000Ps-by	7, 3
Nest 3	24 Jan 2001, 1 egg	26 Jan 2001	–	31 Jan 2001 (failed)	8	–	No	2000_9A	–

FIG. 5.2. Nest and two eggs of *R. melanosticta* in the leaf-sheath of a live *Scheelea* sp. palm tree.

speckling; there was less color at each end. Incubating adults were observed ≤318 m from active nests.

Nestlings.—At least two of the four nesting attempts were successful, and one failed. The fourth nest was found on the last day of a field season, and its outcome is unknown (Table 5.3). One nest fledged two young; another nest fledged only one, though two nestlings were in it until day 11. Although both of the latter nestlings were fully feathered by then, the one that disappeared from the nest did not survive (i.e. it was subsequently never seen with its parents). Its sibling fledged in the morning 13 days after hatching and was measured on the day before it left the nest (Table 5.2). Nestlings from the second successful nest were measured one day before fledging, or at ~12 days old (Table 5.2). Adults with nestlings were observed foraging at antswarms ≤386 m from their active nest.

Fledglings.—All fledglings had a scalloped back similar in pattern to an adult female's. Museum notes from Willis (1969) state that both male and female fledglings have scalloped backs, and males presumably gain the unscalloped adult plumage in their next moult. Fledglings had gray eyeskin and a dark-gray crest. Fledgling 99*Rgw-s* was first seen foraging at an antswarm one month after it fledged. Given that the banded parents had been seen regularly with that ant colony throughout October, that was likely the first day the fledgling left cover and approached the swarm. The adult female aggressively displaced her mate four times in one hour when he approached the fledgling.

Interestingly, the nest of fledgling 99*Rgw-s* was re-used less than two months after 99*Rgw-s* fledged. Because of time constraints, I was unable to determine the identity of the pair attending the two eggs, but a vocalizing female near the nest confirmed the species as *R. melanosticta*. A related antbird of Panama (*G. leucaspis*) is noted to occasionally reuse the same nest twice (Zimmer and Isler 2003). Fledgling 99*Rgw-s* successfully survived to adulthood (confirmed by resight data in later years), so if the pair at the nest was her parents, they were renesting while caring for a dependent fledgling. That behavior has not been documented in other ant-followers.

The second successful nest (00*Rgp-s*) described here fledged two young, which were first seen at an antswarm when they were 13 days out of the nest. The fledglings remained in nearby cover and were fed by their parents. As with *G. salvini*, each parent provisioned both fledglings. Over the next few weeks, fledglings clumsily began foraging on their own. One fledgling began being courted by an adult male when she was less than a month out of the nest. The adult male was accepted by the fledgling's parents and began regularly foraging with them, taking over most duties of provisioning the fledgling. Both fledglings remained with their parents for at least three months, when the field season ended in mid-February. Neither was seen the following year, although the short period of field days (3.5 weeks) or dispersal from the study site may have precluded their being observed.

Phlegopsis nigromaculata

Nest.—Three nests were found in the 2000–2001 field season, of which two fledged young (Table 5.3). If all nests had been successful, nest activity would have spanned October–February. All nests were located in low vertical cavities with completely open tops: two were in stumps of dead palm saplings, and the other was in a live tree (Table 5.1). Nest height above ground ranged from 8 to 51 cm. Cavity depth was from 10 to 15 cm, and cavity diameter was 4.5 to 10 cm. Nests were composed of a few small pieces of dry palm frond, forming a loose flat lining inside each cavity. Nest cavity locations varied in habitat, from gap edge to swamp edge to mature high-ground forest.

Eggs.—No eggs were measured for this species. Egg coloration was similar to that of *R. melanosticta* and *G. salvini*: pale pink with maroon striations. Incubating individuals were observed traveling ≤360 m from their nest to forage at antswarms.

Nestlings.—Two of the three nests each successfully fledged two young (Table 5.3). Young remained in one nest for ~13 days. Four nestlings were measured at ~9 and ~11 days old, respectively (Table 5.2). Botfly larvae parasitized all nestlings; each nestling had one to seven mature larvae at the time of banding (Fig. 5.3). The individual with seven larvae did not show any negative signs in its growth as compared to its sibling and successfully reached adulthood (i.e. was sighted the following year). Adults with nestlings in the nest were observed traveling ≤161 m to forage at antswarms. However, adults were not located on many days, and could have used *L. praedator* swarms of unknown distance from the nest.

Fledglings.—All fledglings were observed at antswarms after fledging, and were still receiving food from adults after 2.5 months out of the nest. Fledglings had black eyeskin, which seemed to slowly turn red over the course of the next many months: in February, older juveniles, presumably from early-rainy-season nests (September–October), had dull reddish eyeskin.

Helpers at the nest.—One nest was observed over the entire nestling stage and confirms for the first time that this species is a cooperative breeder (Table 5.3). Over the course of eight days, we observed the nest for ~18 h and recorded feedings by one adult female and two adult males. At time of nesting, the female (*Ps-by*) was at least two years old, one male (*Pfs-fg*) was at least four, and the second male (*Ps-pg*) was at least two. In 77 observed feedings, the female fed 20% of the time, the older male (*Pfs-fg*) fed 42% of the time, and the younger male (*Ps-pg*) fed 36% of the time. In 2% of the feeding, the adult's identity was unknown. Blood was collected from both nestlings and all three adults, and DNA testing will be carried out to determine paternity and relatedness among all individuals. Too little information exists on cooperative breeding among thamnophilids to speculate as to whether any of the adults involved were related.

Dendrocincla merula

Although the nest of this woodcreeper species remains undescribed, I present information here on breeding biology and fledgling behavior. At Cocha Cashu, nesting appears to begin in the middle of the rainy season, around December. Beginning in January, three females (radiotagged or banded or both) brought one fledgling each to antswarms. No fledglings were ever seen at swarms in the early months of the rainy season (September–December). Fledglings appeared to be almost identical to adults in the field, with a slightly darker bill and a bit less white on the throat area. Fledglings solicited food from their mothers, but also from other *D. merula* females, which either ignored or pecked at them. Fledglings often "played" with each other at swarms rather than foraging. One individual would perch directly behind another on a vertical tree and poke its bill into the back feathers of the other. Both would then fly to a new tree and reverse the "poking order." Fledgling birds would repeat that behavior with up to three nonrelated individuals for ≤10 min at a time, stopping finally to demand food from their respective mothers. At times a fledgling would poke its mother or a nonrelated adult, which would usually fly away to a different perch rather than peck at the fledgling. Willis (1978) describes similar behavior in *D. merula* fledglings in Brazil.

Fledglings stayed with their parents for at least six weeks, but began to become more independent by that time. One individual, at a month post-fledging, was seen alone at an antswarm, 600 m from a second swarm where its mother was foraging. The same individual was observed later that week roosting at least 500 m from where its mother regularly roosted. The fledgling roosted between the buttresses of an emergent *Ficus* sp. tree, 3 m from the ground. Rather than perch on a branch, it held vertically to the smooth bark with its claws and tail, making it virtually impossible for a predator to reach. The next day, mother and fledgling were again foraging together at a swarm.

Although a known nest location is not available, I calculated the minimum distance a

FIG. 5.3. *Phlegopsis nigromaculata* nestlings with larval botflies embedded in skin around neck and head. The first individual had seven large larval botflies at time of measurement; the second (inset) had three.

female traveled from her active nest. Females were identified as having nestlings over a given range of days by counting backward to probable nestling dates from the initial day that a female was observed at a swarm with a fledgling (using information for the genus in Marantz et al. 2004). By examining, *post-hoc*, the distances traveled between swarms by females with active nests and taking the midpoint between swarms, I found that females traveled a minimum of 300 m from the nest while it was active. Given that *D. merula* individuals are strong fliers and sometimes traveled more than a kilometer between successively visited swarms, the actual distance a nesting female may fly is likely much higher.

DISCUSSION

Nests of *G. salvini*, *P. nigromaculata*, and *R. melanosticta* share many attributes, perhaps because of the species' close phylogenetic histories. Male and female of the three species share incubation of eggs and nestlings, and radiotelemetry data taken during the present study show that, at least for *R. melanosticta* and *G. salvini*, the female sits on the nest at night. That is consistent with Skutch's antbird data from Costa Rica (Skutch 1996). In all three species, the nest is placed inside a vertical cavity with a completely open top. The cavity can be a rotting tree stump, a small live tree with a natural cavity formed between branches, or the sheath of a fallen palm frond still attached to a mature *Scheelea* sp. tree. Although sample sizes are small, it seems that *P. nigromaculata* may be restricted to tree cavities and not use *Scheelea* palm sheaths, perhaps because the latter are too small. *Rhegmatorhina melanosticta* is the only species found, thus far, to nest in all three types of cavity. Further study will elucidate whether the smaller *G. salvini* also uses palm-sheath cavities; in Panama, the closely related *G. leucaspis* nests in palm-sheath cavities attached to the trunk of *Scheelea* spp. trees (Willis 1967).

Nest-site selection may contribute to niche breadth among the obligate ant-followers of Cocha Cashu, and may act to limit densities of the dominant *P. nigromaculata* below what the army ant resource could support. If vertical stumps are a limiting resource, smaller species that can successfully nest in *Scheelea* spp. palm sheaths may be able to increase densities to capitalize on the availability of the army ant resource. Brightsmith (2000) experimentally increased availability of nest holes at Cocha Cashu to determine whether secondary nest-holers were limited by low availability of nesting substrates. He found little evidence for that hypothesis, which suggests that *P. nigromaculata* densities may not correlate with nest-site abundance. However, Brightsmith (2000) investigated availability of secondary treeholes, whereas *P. nigromaculata* require open-topped stumps with a narrow range in width and height. Palm stumps are always in a state of decay, and—unlike tree holes—generally do not last for more than one breeding season (S. K. Willson pers. obs.).

Each species constructs a shallow, flat nest of stripped dead palm leaves, about 5–12 cm long × 1 cm wide at the bottom of the cavity. The nest is basically a bed for the eggs and is not cup-shaped or twined together. As a rule, antbirds lay two eggs per nest (Skutch 1985). Eggs of all three species are similarly colored, with a light-pink background and maroon striations or speckling. No nests were observed through the entire egg stage, but one nest was observed with eggs for nine days before they hatched. Skutch (1996) gives the incubation period for the related *G. leucaspis* as 15–16 days. Nestlings of that species remained in nests for 12–13 days, which is a bit shorter than Skutch's (1996) account of 13–15 days.

In southeastern Peru, obligate ant-followers seem to nest in the early to middle rainy season. All nests were found from September to January, and it seems likely that nesting tapers off after January unless nesting attempts repeatedly fail. Juveniles of each species began appearing at antswarms in October, and by January fledglings accompanied many adults. Fledged young of all species described here stayed with their parents for at least three months. If obligate ant-followers must stay with their parents for a protracted period to learn the specialized behavior of following army ants, it seems reasonable that they would not renest if the first nest were successful. However, one *R. melanosticta* pair apparently renested in the same nest less than two months after fledging one nestling. That fledgling successfully survived to adulthood, and was seen in subsequent years. More observations of nest attempts will help answer the question of whether juveniles stay with parents while the latter attempt to rear a second brood.

Mate fidelity and mate switching.—The four antbird study species maintain pair bonds over consecutive breeding seasons. In the course of fieldwork, many banded pairs were identified and observed together for up to three seasons. A total of 37 mated pairs were identified: 13 *R. melanosticta* pairs, 9 *G. salvini*, 9 *M. fortis*, and 6 *P. nigromaculata* (Table 5.4). Those pairs were identified by repeated observation together, mate feedings, nesting attempts, or a combination thereof. *Phlegopsis nigromaculata* pairs were more difficult to determine, because of their propensity to forage in groups and mate cooperatively, and are not represented in Table 5.4. Fourteen known instances of mate switching occurred over four years: seven in the *R. melanosticta* pairs, four in *G. salvini*, two in *M. fortis*, and one in *P. nigromaculata*. In all but one instance, the individual that was replaced from the pair was never seen again and presumably died, which prompted the remaining individual to seek a new mate. Data from 2003 are added to Table 5.4, because the only observed instance of mate switching among living partners occurred in that year (see *R. melanosticta* female s-yp). An individual may quickly find a new mate if its partner dies. For example, a banded *G. salvini* female, which was paired with a banded male for at least two years, paired with a new mate on her home range within two weeks of her first mate's death (the radiotagged male was swallowed by a horned toad [*Ceratophrys cornuta*]). A year later, the female was again with a new male, on a different part of the study plot, ~500 m from the boundary of the home range she shared with her first known mate.

The antbird family is large and currently paraphyletic (Hackett and Rosenberg 1990, Isler et al. 1998, Zimmer and Isler 2003). A re-examination of the thamnophilid phylogeny is underway, and various researchers are using genetic data, vocal characters, and natural-history information in their revisions (Hackett 1993; Isler et al. 1997, 1999; Bates et al. 1999). Within the family Dendrocolaptidae, taxonomic confusion is still widespread at many levels (Marantz et al. 2004). Besides adding to our limited knowledge of the reproduction and natural history of tropical birds, nesting information may provide additional characters for phylogenetic studies (Sheldon and Winkler 1999, Zyskowski and Prum 1999).

TABLE 5.4. Obligate ant-following antbird pairs maintain long-term pair bonds. Individual disappearances may represent deaths or "divorce." One divorce was observed (see *R. melanosticta* female s-yp in 2000–2001 and 2003). No other individuals were seen again after leaving a mate. Mated pairs of *P. nigromaculata* were difficult to discern, and are not listed here. Bird pairs are listed by color-band combinations, with males preceding females. Bold print signifies a change in pair composition; "unb" indicates an unbanded bird. Data are from Cocha Cashu, Peru.

1998	1999	2000–2001	2002	2003
		G. salvini		
yy-s, s-fg	yy-s, **s-oy**	yy-s, s-oy	yy-s, **bw-s**	yy-s, bw-s
	go-s, bw-s	go-s, bw-s		
	pp-s, s-pp	pp-s, s-pp	pp-s, s-pp	**rp-s**, s-pp
	bp-s, s-pb			
		s-yp, s-yp	s-yp, s-yp	s-yp, s-yp
			r-sr, unb	
			s-by, y-sb	unb, rb-s
		R. melanosticta		
s-ww, s-ff	s-ww, s-ff			
pr-s, s-wp	**unb**, s-wp			
by-s, yg-s	by-s, yg-s	by-s, **gp-s**		by-s, **s-yp**
s-rp, s-ow	s-rp, **gp-s**			
	s-yr, s-yp	s-yr, s-yp	s-yr, **wg-s**	s-yr, ?
	s-yg, s-pb			
		r-sr, p-sr		
		s-py, s-gb	s-py, **p-sr**	
		M. fortis		
ob-s, rw-s		ob-s, **g-sy**	ob-s, **rr-s**	**unb**, rr-s
x-s, s-bw	x-s, s-bw			
	s-rg, s-pp	s-rg, s-pp		
		s-yr, ry-s		
		s-yy, ff-s	s-yy, ff-s	s-yy, **unb**
			s-bb, bb-s	s-bb, **unb**
			r-sg, s-gr	

6. CONCLUSIONS AND FUTURE QUESTIONS

Summary of Coexistence Patterns

The present study contributes new insight toward understanding the complexities of community organization—a major goal of the study of ecology. The five obligate ant-following birds at Cocha Cashu coexist through more subtle means than segregation by simple ecological factors like body size, habitat, or prey size. Two of the five species (*D. merula* and *M. fortis*) preferred one or the other of the two army ant species as a foraging resource. The woodcreeper *D. merula* further segregated from the four antbirds in its use of white-lipped peccary herds—which act as beaters of arthropod prey in a manner similar to that of the army ants—as a foraging resource. The three antbirds that did not prefer one army ant species over the other (*P. nigromaculata, R. melanosticta,* and *G. salvini*) are segregated by body mass, which may allow differential use of space along a swarm front's width. That would permit a smaller, more subordinate species to "fit" along the front of a swarm that was already "full" to a larger species. These factors add to descriptions of niche differentiation by perch type and angle published by Willis (1968, 1969, 1978, 1979, 1985).

The bird species' population dynamics were not stable over five years of data collection, and the total population of obligate ant-followers dropped by almost half over the course of the study. I suggest that periodic population fluctuations are a normal occurrence in obligate ant-following guilds. Declines over the course of the study were partially attributable to changes in availability of army ants during forest flooding, which may have decreased adult survival in *R. melanosticta*. Because I did not study the birds in the dry seasons, it is unknown whether dry-season events contributed to mortality. On BCI, loss of *Pha. mcleannani* was attributed to a long 1968 drought that caused a short breeding season and lack of replacements due to mesopredation (Willis 1973, 1974). Declines at Cocha Cashu were also attributable to a low juvenile recruitment rate that failed to offset normal adult mortality. Nest-site availability may contribute to fluctuations in recruitment, because suitable stumps and palm sheaths will generally biodegrade and fall apart after a breeding season. Not only are the food resources of obligate ant-followers continually moving across the landscape; nest sites are also changing seasonally as older palm stumps degrade, leaves fall from *Scheelea* spp. trunks, and small palms die and create new stumps.

Population fluctuations decreased interference competition among individuals and may increase the ability of the subordinate species *R. melanosticta* to coexist with the larger, dominant *P. nigromaculata* at Cocha Cashu. In terms of army ant use patterns, *R. melanosticta*'s niche space was not well differentiated from either that of the larger *P. nigromaculata* or the smaller *G. salvini*. I suggest that *R. melanosticta,* in particular, may rely on competitive release resulting from population fluctuations of the dominant species, as well as fewer constraints on nest-site selection, to coexist in the Cocha Cashu bird assemblage.

Few species of tropical forest birds have been intensively studied. The present research—focusing on a guild of highly specialized "atypical" birds—provides important insight into hypotheses concerning the stability of more "normal" (i.e. generalist, territorial, insectivorous) tropical species; the results support the hypothesis that avian territoriality contributes to the stability of tropical bird populations (Greenberg and Gradwohl 1986, Newton 1998). All three widely ranging obligate antbird species (*P. nigromaculata, R. melanosticta,* and *G. salvini*) exhibited significant changes in population levels over three years, whereas the territorial antbird species (*M. fortis*) displayed a constant population density. Adult survival rates in most years were constant across species, which suggests that juvenile recruitment is the key difference, and is either lower or more volatile in nonterritorial bird species than in related territorial species. That prediction deserves further testing in both obligate ant-followers and other bird guilds with variable territorial behavior, while controlling for environmental, temporal, and landscape effects. The population fluctuations of the widely ranging study species, documented here, provide a counterpoint to a number of studies that have found stable population dynamics in territorial tropical birds. Some key ecological differences that I found among the five obligate ant-following species are summarized in Table 6.1.

Nest-site selection contributes to niche breadth among the obligate ant-followers, and may act to constrain population density of the dominant species *P. nigromaculata*. *Dendrocincla merula*, a woodcreeper, presumably nests in tree holes with a vertical opening, like other woodcreepers. The antbird *M. fortis* differentiates from the other obligate antbirds in its construction of terrestrial dome nests. Phylogenetics likely contribute to that differentiation by constraining the nest-substrate choices of the three more closely related antbirds *P. nigromaculata, R. melanosticta*, and *G. salvini*, all three of which nest in open-topped vertical stumps. However, *R. melanosticta* also nests in *Scheelea* spp. palm sheaths, which act as vertical cavities. That tree genus is common throughout the Neotropics; Cocha Cashu has 25 adult trees ha^{-1} (J. Terborgh pers. comm.). High availability of palm sheaths for nesting would remove a limiting factor on bird species that could profit by using them. Presumably, *G. salvini* also uses palm sheaths, as has been demonstrated for its congener *G. leucaspis* in Panama (Zimmer and Isler 2003). *Scheelea* palm sheaths are small and likely unsuitable for the dominant species *P. nigromaculata*. Vertical open-topped wood stumps may thus act as a key limiting factor that constrains local populations of *P. nigromaculata* from expanding to use more of the available army ant resource. Constraints on the dominant species may provide the socially subordinate *R. melanosticta* and *G. salvini* with the niche space they need to coexist with it.

Future Questions

The work presented here opens the door for new questions about the ecology and behavior of obligate ant-following species; the interactions of specialization and vulnerability, tropical stability, and territoriality; and the potential role of source–sink dynamics between contiguous floodplain and *terra firme* habitats in the Amazon basin. All data presented here were collected in the rainy

Table 6.1. Summary of some key behavioral and ecological differences among five species of obligate ant-following bird species at Cocha Cashu Biological Station, Manu National Park, Peru.

Species	Mass (g)	Rank in dominance hierarchy	Mating system	Army ant species preference	Mean adult survival rate	Mean recruitment per year per 100 ha	Mean home-range size(ha)	Defended or undefended area	Mean swarm-front use (m)	Nest substrate
P. nigromaculata	46.0	1	Cooperative	Both	0.63	1.1	46	Undefended home range	1.06	Stump
M. fortis	46.5	2	Paired	*L. praedator*	0.72	2.3	15	Defended territory	1.16	Dome nest on ground
R. melanosticta	31.4	3	Paired	Both	0.56	0.5	48	Undefended home range	0.73	Stump, *Scheelea* sheath
D. merula	47.5	4	Solitary	*E. burchelli*	0.80	0.6	65	Undefended home range	1.40	Tree hole
G. salvini	25.9	5	Paired	Both	0.59	1.7	27	Undefended home range	0.71	Stump, probably *Scheelea* sheath

season, and it is not known how the army ant and obligate ant-follower system changes in the dry season. Data from BCI (Willis 1967; Franks 1982a, b) suggest that *E. burchelli* densities remain stable through seasonal changes. However, *L. praedator* swarms may be less available to birds in the dry season, when ants may forage more often below ground or at night (Schneirla 1971, Hölldobler and Wilson 1990). How does that affect the territorial system of *M. fortis*? At the constant density that I estimated during my study, an increase in territory size would force pairs to have some overlap in neighboring territories. A comparison of *M. fortis* behavior and spatial dynamics across seasons would provide insight into the effects of temporal resource availability on territorial plasticity. Behavioral plasticity, as well, may occur if *M. fortis* cannot afford its conspecific intolerance because of larger territory size in the dry season.

A second question, of larger scope, is how territoriality contributes to the stability of tropical bird populations. In the obligate ant-followers, data presented here are consistent with the hypothesis that nonterritorial birds have larger population fluctuations (both positive and negative) than territorial species. Most tropical bird work has concentrated on territorial species; an investigation of this question will profit from collection of long-term population data on various tropical guilds that include both territorial and nonterritorial species.

My proposed hypothesis that nest-site availability is a key limiting factor for populations of *P. nigromaculata* needs to be tested. Nest-site limitation may act to depress populations of this dominant species below the environment's carrying capacity based on food-resource availability. Because they use palm sheaths, the subordinate species *R. melanosticta* and *G. salvini* are likely less constrained in nest substrate and would be able to utilize some of the space at antswarms beyond what the *P. nigromaculata* population can occupy. Nest-site limitation may also explain another phenomenon specific to *P. nigromaculata*—cooperative breeding. If nesting stumps are so limiting that there are few benefits to younger birds in leaving the natal territory to initiate pairing and nesting, that would imply a potential cause of the evolution of cooperative breeding in the species. An experimentally expanded selection of nest stumps may allow for population increase in *P. nigromaculata*, over time. However, the question of how stump availability affects the breeding behavior of the species would be easier to answer in a short-term study. If "helpers at the nest" are constrained by nest availability, we might predict that helpers disperse when nest sites are made available. Results presented here from one nest suggest that males may breed cooperatively with a single female, but too little information is known on *P. nigromaculata* reproductive biology to discount female helpers as well. Finally, groups of *P. nigromaculata* may form for reasons other than nesting constraints. Do larger groups repel smaller conspecific groups from swarms? Further study can suggest answers to these questions.

Are floodplain forests in the Amazon a sink or a source habitat for obligate ant-following birds? Floodplain habitat constitutes only 3% to 4% of the Amazon basin, yet the ubiquitous system of rivers running like veins throughout the Amazon means that no region is far from floodplain (Goulding 1993). For obligate ant-followers, floodplain may offer higher food-resource density, but at the potential cost of higher seasonal variability. Floodplain variability and ensuing changes in resource availability may cause alterations in relative abundances of bird species in different years, as was seen in the five study species. In good years, juveniles may disperse into *terra firme* forest. Floodplain may act as a source for the *terra firme*, which should hold lower populations of insectivores because of lower moisture and thus lower arthropod abundance (Williams 1941, Janzen and Schoener 1968, Janzen 1973, Wolda 1978, Levings and Windsor 1982). However, bad years for some species in the floodplain may translate into the *terra firme* acting as a source, as individuals immigrate into the floodplain and raise populations of those species back to carrying capacity. Long-term studies of dispersal of juveniles from adjoining habitats, in conjunction with measures of resource density in both habitats, are necessary to answer this question. Source–sink dynamics are well established in temperate regions, where fragmentation, predation, and nest parasitism clearly define what areas act as sources versus sinks (Donovan et al. 1995, Robinson et al. 1995, Thompson et al. 2002). In the contiguous lowland rainforest of the Amazon, the natural dynamics of floodplain ecosystems and the greater stability of *terra firme* forests may dictate that sources and sinks vary year to year, as populations rise and fall because of fluctuations in rainfall levels, prey availability, competitive pressures, predation, and ensuing survival and recruitment.

Conservation Applications

The natural world is rapidly being degraded and fragmented by human activities. My research, done in a protected area of the Amazon, provides crucial baseline data on population dynamics and densities, survival rates, and home-range requirements of obligate ant-following birds. It also provides baseline ecological data for the two army ant species, *E. burchelli* and *L. praedator*—data were not previously available for Amazonian populations. Army ants and obligate ant-following birds are among the first groups of organisms to go locally extinct in fragmented forest (Harper 1987, 1989; Bierregaard and Lovejoy 1989; Stouffer and Bierregaard 1995; Bierregaard and Stouffer 1997). Presence or absence of these species in fragmented or threatened forest may be an excellent indicator of forest degradation or health (Thiollay 1992, Canaday 1997). Both the army ants and obligate ant-following birds are potential indicator species for use in rapid ecological assessments of threatened areas.

ACKNOWLEDGMENTS

I thank my dissertation committee members—John Faaborg, Rex Cocroft, Bette Loiselle, John Terborgh, and Frank Thompson III—for their interest and insight, and for invaluable suggestions that improved this manuscript. The work presented here would have been impossible without the competent assistance of the following field assistants: Luz Jimenez, Lindy McBride, Gwylim Blackburn, Fred Werner, Daniel Huaman, C. E. Timothy Paine, David Wilamowski, Rich Pagen, Dan Hogan, Kevin Svara, Edith Suazo, Stuart Hillman, Silvia Castro Delgado, Stacy Small, James DeStaebler, Mike Stake, Wendy Jess, Kristin Wornson, Windy Davis, and Nadia Castro Izaguirre. Gordon Burleigh and Tara Robinson carried out sex determination of monomorphic bird species. Discussions with Mercedes Foster, Katie Dugger, Jeff Rouder, and Brett Sandercock substantially improved the manuscript. Special thanks to my husband, Stuart Hillman, who helped in many ways. Funding for this study was provided by a Teagle Foundation Scholarship, a Graduate Assistance in Areas of National Need (GAANN) Fellowship from the U.S. Department of Education, two Trans World Airlines Environmental Scholarships, an Organization of American States Regular Training Program (PRA) Graduate Fellowship, and travel grants from the Division of Biological Sciences at the University of Missouri and from the North American Ornithological Conference. I thank the Instituto Nacional de Recursos Naturales (INRENA) of Peru for permission to live and work at Cocha Cashu Biological Station between 1997 and 2002. I would like to acknowledge a friend from Cocha Cashu, the late Francis Bossuyt, whose dedication to his work in the field awed and inspired me. I would also like to thank the Matsigenka people for their respect of the no-hunting boundaries around Cocha Cashu, which make it such a special place.

LITERATURE CITED

ABRAMS, P. 1983. The theory of limiting similarity. Annual Review of Ecology and Systematics 14: 359–376.

BATES, H. W. 1862. The Naturalist on the River Amazons. John Murray, London.

BATES, J. M., S. J. HACKETT, AND J. M. GOERCK. 1999. High levels of mitochondrial DNA differentiation in two lineages of antbirds (*Drymophila* and *Hypocnemis*). Auk 116:1093–1106.

BIERREGAARD, R. O., JR., AND T. E. LOVEJOY. 1989. Effects of forest fragmentation on Amazonian understory bird communities. Acta Amazônica 19:215–241.

BIERREGAARD, R. O., JR., T. E. LOVEJOY, C. GASCON, AND R. MESQUITA, EDS. 2002. Lessons from Amazonia: The Ecology and Conservation of a Fragmented Forest. Yale University Press, New Haven, Connecticut.

BIERREGAARD, R. O., JR., AND P. C. STOUFFER. 1997. Understory birds and dynamic habitat mosaics in Amazonian rainforests. Pages 138–155 *in* Tropical Forest Remnants: Ecology, Management, and Conservation of Fragmented Communities (W. F. Laurance and R. O. Bierregaard, Jr., Eds.). University of Chicago Press, Chicago.

BOWDEN, D. C. 1993. A Simple Technique for Estimating Population Size. Department of Statistics, Colorado State University, Fort Collins.

BRIGHTSMITH D. J. 2000. Use of arboreal termitaria by nesting birds in the Peruvian Amazon. Condor 102:529–538.

BROWN, J. H. 1975. Geographical ecology of desert rodents. Pages 315–341 *in* Ecology and Evolution of Communities (M. L. Cody and J. M. Diamond, Eds.). Harvard University Press, Cambridge, Massachusetts.

BROWN, J. L. 1964. The evolution of diversity in avian territorial systems. Wilson Bulletin 76:160–169.

BROWN, J. S. 1989a. Coexistence on a seasonal resource. American Naturalist 133:168–182.

BROWN, J. S. 1989b. Desert rodent community structure: A test of four mechanisms of coexistence. Ecological Monographs 59:1–20.

BURNHAM, K. P., AND D. R. ANDERSON. 2002. Model Selection and Multimodel Inference: A Practical Information–Theoretic Approach. Springer-Verlag, New York.

CADENA, C. D., G. A. LONDOÑO, AND J. L. PARRA. 2000. Nesting records of five antbird species from the Colombian Amazon. Wilson Bulletin 112:313–317.

CANADAY, C. 1997. Loss of insectivorous birds along a gradient of human impact in Amazonia. Biological Conservation 77:63–77.

CHARNOV, E. L. 1976. Optimal foraging: The marginal value theorem. Theoretical Population Biology 9:129–136.

CHESSER, R. T. 1995. Comparative diets of obligate ant-following birds at a site in northern Bolivia. Biotropica 27:382–390.

CODY, M. L. 1966. A general theory of clutch size. Evolution 20:174–184.

COEHLO, C. A. S., C. B. UVO, AND T. AMBRIZZI. 2002. Exploring the impacts of tropical Pacific SST on the precipitation patterns over South America during ENSO periods. Theoretical and Applied Climatology 71:185–197.

CONDIT, R., S. P. HUBBELL, AND R. B. FOSTER. 1996. Changes in tree species abundance in a Neotropical forest over eight years: Impact of climate change. Journal of Tropical Ecology 12: 231–256.

CONNELL, J. H. 1961. The influence of interspecific competition and other factors on the distribution of the barnacle *Chthamalus stellatus*. Ecology 42:710–723.

CONNELL, J. H. 1978. Diversity in tropical rain forests and coral reefs. Science 199:1302–1310.

CONNELL, J. H., AND E. ORIAS. 1964. The ecological regulation of species diversity. American Naturalist 98:399–414.

DEL HOYO, J., A. ELLIOTT, AND D. A. CHRISTIE, EDS. 2004. Handbook of the Birds of the World, vol. 8. Lynx Edicions, Barcelona, Spain.

DIAMOND, J. M. 1975. Assembly of species communities. Pages 342–444 *in* Ecology and Evolution of Communities (M. L. Cody and J. M. Diamond, Eds.). Belknap Press of Harvard University Press, Cambridge, Massachusetts.

DOBZHANSKY, T. 1950. Evolution in the tropics. American Scientist 38:209–221.

DONOVAN, T. M., R. H. LAMBERSON, F. R. THOMPSON III, AND J. FAABORG. 1995. Modeling the effects of habitat fragmentation on source and sink demography of Neotropical migrant birds. Conservation Biology 9:1396–1407.

DURANT, S. M. 1998. Competition refuges and coexistence: An example from Serengeti carnivores. Journal of Animal Ecology 67:370–386.

EMMONS, L. H. 1990. Neotropical Rainforest Mammals: A Field Guide. University of Chicago Press, Chicago.

FAABORG, J. 1982. Avian population fluctuations during drought conditions in Puerto Rico. Wilson Bulletin 94:20–30.

FEDOROV, A. A. 1966. The structure of the tropical rain forest and speciation in the humid tropics. Journal of Ecology 54:1–11.

FEINSINGER, P. 1976. Organization of a tropical

guild of nectarivorous birds. Ecological Monographs 46:257–291.

Fischer, A. G. 1960. Latitudinal variation in organic diversity. Evolution 14:64–81.

Fitzpatrick, J. W. 1985. Form, foraging behavior, and adaptive radiation in the Tyrannidae. Ornithological Monographs 36: 447–470.

Fogden, M. P. L. 1972. The seasonality and population dynamics of equatorial forest birds in Sarawak. Ibis 114:307–343.

Forsyth, A., and K. Miyata. 1984. Tropical Nature. Charles Scribner's Sons, Macmillan Publishing, New York.

Franks, N. R. 1980. The evolutionary ecology of the army ant *Eciton burchelli* on Barro Colorado Island, Panama. Ph.D. dissertation, University of Leeds, Leeds, United Kingdom.

Franks, N. R. 1982a. A new method for censusing animal populations: The number of *Eciton burchelli* army ant colonies on Barro Colorado Island, Panama. Oecologia 52:266–268.

Franks, N. R. 1982b. Ecology and population regulation in the army ant *Eciton burchelli*. Pages 389–395 *in* The Ecology of a Tropical Forest (E. G. Leigh, Jr., A. S. Rand, and D. M. Windsor, Eds.). Smithsonian Institution Press, Washington, D.C.

Fretwell, S. D., and H. L. Lucas. 1970. On territorial behaviour and other factors influencing habitat distribution in birds. Acta Biotheoretica 19:16–36.

Gaston, K. J., Ed. 1996. Biodiversity: A Biology of Number and Difference. Blackwell Scientific Publications, Oxford.

Gause, G. F. 1934. The Struggle for Existence. Williams and Wilkins, Baltimore, Maryland.

Gentry, A. H., Ed. 1990. Four Neotropical Rainforests. Yale University Press, New Haven, Connecticut.

Goulding, M. 1993. Flooded forests of the Amazon. Scientific American 266:114–120.

Grant, P. R. 1985. Climatic fluctuations on the Galapagos Islands and their influence on Darwin's finches. Ornithological Monographs 36:471–483.

Greenberg, R., and J. Gradwohl. 1986. Constant density and stable territoriality in some tropical insectivorous birds. Oecologia 69:618–625.

Greenberg, R., and J. Gradwohl. 1997. Territoriality, adult survival, and dispersal in the Checker-throated Antwren in Panama. Journal of Avian Biology 28:103–110.

Griffiths, R., M. C. Double, K. Orr, and R. J. G. Dawson. 1998. A DNA test to sex most birds. Molecular Ecology 7:1071–1075.

Grinnell, J. 1922. The trend of avian populations in California. Science 56:671–676.

Hackett, S. J. 1993. Phylogenetic and biogeographic relationships in the Neotropical genus *Gymnopithys* (Formicariidae). Wilson Bulletin 105:301–315.

Hackett, S. J., and K. V. Rosenberg. 1990. Comparison of phenotypic and genetic differentiation in South American antwrens (Formicariidae). Auk 107:473–489.

Harper, L. H. 1987. The conservation of ant-following birds in small Amazonian forest fragments. Ph.D dissertation, State University of New York, Albany.

Harper, L. H. 1989. The persistence of ant-following birds in small Amazonian forest fragments. Acta Amazônica 19:249–263.

Himes, J. G. 2003. Intra- and interspecific competition among the water snakes *Nerodia sipedon* and *Nerodia rhombifer*. Journal of Herpetology 37:126–131.

Holdridge, L. R. 1967. Life Zone Ecology. Centro de Ciencias Tropicales, San José, Costa Rica.

Hölldobler, B., and E. O. Wilson. 1990. The Ants. Harvard University Press, Cambridge, Massachusetts.

Holmgren, M., M. Scheffer, E. Ezcurra, J. R. Gutiérrez, and G. M. J. Mohren. 2001. El Niño effects on the dynamics of terrestrial ecosystems. Trends in Ecology and Evolution 16:89–94.

Hooge, P. N., and B. Eichenlaub. 1997. Animal movement extension to ARCVIEW, version 1.1. Alaska Science Center–Biological Science Office, U.S. Geological Survey, Anchorage, Alaska.

Huston, M. A. 1979. A general hypothesis of species diversity. American Naturalist 113:81–101.

Hutchinson, G. E. 1957. Concluding remarks. Population Studies: Animal Ecology and Demography. Cold Spring Harbor Symposia on Quantitative Biology 22:415–427.

Hutchinson, G. E. 1959. Homage to Santa Rosalia or why are there so many kinds of animals? American Naturalist 93:145–159.

Isler, M. L., P. R. Isler, and B. M. Whitney. 1997. Biogeography and systematics of the *Thamnophilus punctatus* (Thamnophilidae) complex. Ornithological Monographs 48: 355–381.

Isler, M. L., P. R. Isler, and B. M. Whitney. 1998. Use of vocalizations to establish species limits in antbirds (Thamnophilidae). Auk 115: 577–590.

Isler, M. L., P. R. Isler, and B. M. Whitney. 1999. Species limits in antbirds (Passeriformes: Thamnophilidae): The *Myrmotherula surinamensis* complex. Auk 116:83–96.

Ives, A. R. 1995. Measuring competition in a spatially heterogeneous environment. American Naturalist 146:911–936.

Janzen, D. H. 1967. Why mountain passes are higher in the tropics. American Naturalist 101: 233–249.

Janzen, D. H. 1973. Sweep samples of tropical foliage insects: Effects of seasons, vegetation types, elevation, time of day, and insularity. Ecology 54:687–708.

Janzen, D. H., and T. W. Schoener. 1968. Differences in insect abundance and diversity between wetter and drier sites during a tropical dry season. Ecology 49:96–110.

Jullien, M., and J.-M. Thiollay. 1998. Multi-species territoriality and dynamics in Neotropical forest understorey bird flocks. Journal of Animal Ecology 67:227–252.

Karr, J. R. 1971. Structure of avian communities in selected Panama and Illinois habitats. Ecological Monographs 41:207–233.

Karr, J. R. 1989. Birds. Pages 401–416 *in* Tropical Rain Forest Ecosystems (H. Leith and M. J. A. Werger, Eds.). Elsevier Science, Amsterdam, The Netherlands.

Karr, J. R., and K. E. Freemark. 1983. Habitat selection and environmental gradients: Dynamics in the "stable" tropics. Ecology 64: 1481–1494.

Karr, J. R., D. W. Schemske, and N. V. L. Brokaw. 1982. Temporal variation in the understory bird community of a tropical forest. Pages 441–453 *in* The Ecology of a Tropical Forest (E. G. Leigh, Jr., A. S. Rand, D. M. Windsor, Eds.). Smithsonian Institution Press, Washington, D.C.

Kricher, J. 1997. A Neotropical Companion, 2nd ed. Princeton University Press, Princeton, New Jersey.

Lack, D. 1944. Symposium on the ecology of closely allied species. Journal of Animal Ecology 13: 176–177.

Lebreton, J.-D., K. P. Burnham, J. Clobert, and D. R. Anderson. 1992. Modeling survival and testing biological hypotheses using marked animals: A unified approach with case studies. Ecological Monographs 62:67–118.

Leigh, E. G., Jr. 1982. Introduction: The significance of population fluctuations. Pages 435–440 *in* The Ecology of a Tropical Forest (E. G. Leigh, Jr., A. S. Rand, and D. M. Windsor, Eds.). Smithsonian Institution Press, Washington, D.C.

Leigh, E. G., Jr. 1999. Tropical Forest Ecology. Oxford University Press, New York.

Leigh, E. G., Jr., and D. M. Windsor. 1982. Forest production and regulation of primary consumers on Barro Colorado Island. Pages 111–122 *in* The Ecology of a Tropical Forest (E. G. Leigh, Jr., A. S. Rand, and D. M. Windsor, Eds.). Smithsonian Institution Press, Washington, D.C.

Levey, D. J., and C. Martínez del Rio. 2001. It takes guts (and more) to eat fruit: Lessons from avian nutritional ecology. Auk 118:819–831.

Levings, S. C., and D. M. Windsor. 1982. Seasonal and annual variation in litter arthropod populations. Pages 355–387 *in* The Ecology of a Tropical Forest (E. G. Leigh, Jr., A. S. Rand, and D. M. Windsor, Eds.). Smithsonian Institution Press, Washington, D.C.

Loiselle, B. A., and W. G. Hoppes. 1985. Nest predation in insular and mainland lowland rainforest in Panama. Condor 85:93–95.

MacArthur, R. H. 1958. Population ecology of some warblers of northeastern coniferous forests. Ecology 39:599–619.

MacArthur, R. H. 1969. Patterns of communities in the tropics. Biological Journal of the Linnean Society 1:19–30.

MacArthur, R. H. 1972. Geographical Ecology: Patterns in the Distribution of Species. Harper and Row, New York.

MacArthur R. H., and R. Levins. 1964. Competition, habitat selection, and character displacement in a patchy environment. Proceedings of the National Academy of Sciences USA 51:1207–1210.

MacArthur, R. H., and R. Levins. 1967. The limiting similarity, convergence and divergence of coexisting species. American Naturalist 101: 377–385.

Marantz, C. A., A. Aleixo, L. R. Bevier, and M. A. Patten. 2003. Family Dendrocolaptidae (Woodcreepers). Pages 358–447 *in* Handbook of the Birds of the World, vol. 8. (J. del Hoyo, A. Elliott, and D. A. Christie, Eds.). Lynx Edicions, Barcelona, Spain.

Marengo, J. A. 1992. Interannual variability of surface climate in the Amazon basin. International Journal of Climatology 12:853–863.

Marra, P. P., and J. V. Remsen, Jr. 1997. Insight into the maintenance of high species diversity in the Neotropics: Habitat selection and foraging behavior in understory birds of tropical and temperate forests. Ornithological Monographs 48:445–483.

May, R. 1973. Stability and Complexity in Model Ecosystems. Princeton University Press, Princeton, New Jersey.

McNair, J. N. 1982. Optimal giving up times and the marginal value theorem. American Naturalist 119:511–529.

Morton, E. S., and B. J. M. Stutchbury. 2000. Demography and reproductive success in the dusky antbird, a sedentary tropical passerine. Journal of Field Ornithology 71:493–500.

Munn, C. A. 1985. Permanent canopy and understory flocks in Amazonia: Species composition and population density. Ornithological Monographs 36:683–712.

Munn, C. A., and J. Terborgh. 1979. Multi-species territoriality in Neotropical foraging flocks. Condor 81:338–347.

Nams, V. O. 2000. Program Locate II. Pacer, Truro, Nova Scotia, Canada. [Online.] Available at www.nsac.ns.ca/envsci/staff/vnams/Locate.htm

Newton, I. 1998. Population Limitation in Birds. Academic Press, New York.

North American Banding Council. 2001a. The Instructor's Guide to Training Passerine Bird Banders in North America. [Online.] North American Banding Council, Point Reyes Station, California. Available at nabanding.net/nabanding/

North American Banding Council. 2001b. The North American Banders' Manual for Banding Passerines and Near Passerines (Excluding Hummingbirds and Owls). [Online.] North American Banding Council, Point Reyes Station, California. Available at nabanding.net/nabanding/

North American Banding Council. 2001c. The North American Banders' Study Guide. [Online.] North American Banding Council, Point Reyes Station, California. Available at nabanding.net/nabanding/

Orians, G. H. 1969. The number of bird species in some topical forests. Ecology 50:783–801.

Pearson, D. L., and J. A. Derr. 1986. Seasonal patterns in lowland forest floor arthropod abundance in southeastern Peru. Biotropica 18:244–256.

Pezzi, L. P., and I. F. A. Cavalcanti. 2001. The relative importance of ENSO and tropical Atlantic sea surface temperature anomalies for seasonal precipitation over South America: A numerical study. Climate Dynamics 17:205–212.

Pianka, E. R. 1966. Latitudinal gradients in species diversity: A review of concepts. American Naturalist 100:33–46.

Pianka, E. R. 1981. Competition and niche theory. Pages 167–196 *in* Theoretical Ecology (R. M. May, Ed.). Sinauer Associates, Sunderland, Massachusetts.

Pierpont, N. 1986. Interspecific aggression and the ecology of woodcreepers (Aves: Dendrocolaptidae). Ph.D. dissertation, Princeton University, Princeton, New Jersey.

Pulliam, H. R., and B. J. Danielson. 1991. Sources, sinks, and habitat selection: A landscape perspective on population dynamics. American Naturalist 137:S50–S66.

Raim, A. 1978. A radio transmitter attachment for small passerines. Bird-Banding 49:326–332.

Rappole, J. H., and A. R. Tipton. 1991. New harness design for attachment of radio transmitters to small passerines. Journal of Field Ornithology 62:335–337.

Rettenmeyer, C. W. 1963. Behavioral studies of army ants. University of Kansas Scientific Bulletin 44:281–465.

Ricklefs, R. E., and D. Schluter, Eds. 1993. Species Diversity in Ecological Communities: Historical and Geographical Perspectives. University of Chicago Press, Chicago.

Ridgely, R. S., and P. J. Greenfield. 2001. The Birds of Ecuador, vol. 2. Cornell University Press, Ithaca, New York.

Robinson, S. K., and J. Terborgh. 1995. Interspecific aggression and habitat selection by Amazonian birds. Journal of Animal Ecology 64:1–11.

Robinson, S. K., F. R. Thompson III, T. M. Donovan, D. R. Whitehead, and J. Faaborg. 1995. Regional forest fragmentation and the nesting success of migratory birds. Science 267:1987–1990.

Robinson, T. R. 2000. Factors affecting natal dispersal by Song Wrens (*Cyphorhinus phaeocephalus*): Ecological constraints and demography. Ph.D. dissertation, University of Illinois, Urbana-Champaign.

Robinson, W. D. 1999. Long-term changes in the avifauna of Barro Colorado Island, Panama, a tropical forest isolate. Conservation Biology 13:85–97.

Robinson, W. D., J. D. Brawn, and S. K. Robinson. 2000a. Forest bird community structure in central Panama: Influence of spatial scale and biogeography. Ecological Monographs 70: 209–235.

Robinson, W. D., T. R. Robinson, S. K. Robinson, and J. D. Brawn. 2000b. Nesting success of understory forest birds in central Panama. Journal of Avian Biology 31:151–164.

Roper, J. J. 1996. Nest predation and its importance for a Neotropical bird, the Western Slaty Antshrike (*Thamnophilus atrinucha*): The problem, experiments, and simulations. Ph.D. dissertation, University of Pennsylvania, Philadelphia.

Rosenberg, K. V. 1997. Ecology of dead-leaf foraging specialists and their contribution to Amazonian bird diversity. Ornithological Monographs 48:673–700.

Rosenzweig, M. L. 1995. Species Diversity in Space and Time. Cambridge University Press, Cambridge, United Kingdom.

SAS Institute. 1997. SAS Software. SAS Institute, Cary, North Carolina.

Schneirla, T. C. 1971. Army Ants: A Study in Social Organization. W. H. Freeman, San Francisco, California.

Schoener, T. W. 1968. Sizes of feeding territories among birds. Ecology 49:123–141.

Schoener, T. W. 1974. Resource partitioning in ecological communities. Science 185:27–39.

Seaman, D. E., J. J. Millspaugh, B. J. Keernohan,

G. C. Brundige, K. J. Raedeke, and R. A. Gitzen. 1999. Effects of sample size on kernel home range estimates. Journal of Wildlife Management 63:739–747.

Seaman, D. E., and R. A. Powell. 1996. An evaluation of the accuracy of kernel density estimators for home range analysis. Ecology 77: 2075–2085.

Sheldon, F. H., and D. W. Winkler. 1999. Nest architecture and avian systematics. Auk 116: 875–877.

Sherry, T. W. 1984. Comparative dietary ecology of sympatric, insectivorous Neotropical flycatchers (Tyrannidae). Ecological Monographs 54: 313–338.

Sieving, K. E. 1992. Nest predation and differential insular extinction among selected forest birds of central Panama. Ecology 73:2310–2328.

Silman, M. R., J. W. Terborgh, and R. A. Kiltie. 2003. Population regulation of a dominant rain forest tree by a major seed predator. Ecology 84:431–438.

Skutch, A. F. 1985. Clutch size, nesting success, and predation on nests of Neotropical birds, revisited. Ornithological Monographs 36:575–594.

Skutch, A. F. 1996. Antbirds and Ovenbirds. University of Texas Press, Austin.

Smythe, N. 1974. Biological monitoring data—Insects. Pages 70–115 *in* 1973 Environmental Monitoring and Baseline Data (R. W. Rubinoff, Ed.). Smithsonian Institution Environmental Science Program, Washington, D.C.

Snedecor, G. W., and W. G. Cochran. 1989. Statistical Methods. Iowa State University Press, Ames.

Spruce, R. 1908. Notes of a Botanist on the Amazon and Andes. MacMillan, London.

Stiles, F. G. 1992. Effects of a severe drought on the population biology of a tropical hummingbird. Ecology 73:1375–1390.

Stouffer, P. C., and R. O. Bierregaard, Jr. 1993. Spatial and temporal abundance patterns of Ruddy Quail-Doves near Manaus, Brazil. Condor 95:896–903.

Stouffer, P. C., and R. O. Bierregaard. 1995. Use of Amazonian forest fragments by understory insectivorous birds. Ecology 76:2429–2445.

Stouffer, P. C., and R. O. Bierregaard. 1996. Seasonal variation in hummingbird abundance in a fragmented Amazonian landscape. Ararajuba (Journal of the Brazilian Ornithological Society) 4:9–14.

Stutchbury, B. J. M., and E. S. Morton. 2001. Behavioral Ecology of Tropical Birds. Academic Press, New York.

Styrsky, J. N. 2003. Life-history evolution and population dynamics of a Neotropical forest bird (*Hylophylax naevioides*). Ph.D. dissertation, University of Illinois, Urbana-Champaign.

Sutherland, W. J., and G. A. Parker. 1985. Distribution of unequal competitors. Pages 255–274 *in* Behavioural Ecology: Ecological Consequences of Adaptive Behaviour (R. M. Sibly and R. H. Smith, Eds.). Blackwell Scientific Publications, Oxford, United Kingdom.

Swartz, M. B. 2001. Bivouac checking, a novel behavior distinguishing obligate from opportunistic species of army-ant-following birds. Condor 103:629–632.

Terborgh, J. 1980. Vertical stratification of a Neotropical forest bird community. Pages 1005–1012 *in* Acta XVII Congressus Internationalis Ornithologici (R. Nöhring, Ed.). Deutsche Ornithologen-Gesellschaft, Berlin.

Terborgh, J. 1983. Five New World Primates: A Study in Comparative Ecology. Princeton University Press, Princeton, New Jersey.

Terborgh, J. 1990. An overview of research at Cocha Cashu Biological Station. Pages 48–59 *in* Four Neotropical Forests (A. H. Gentry, Ed.). Yale University Press, New Haven, Connecticut.

Terborgh, J. W., J. W. Fitzpatrick, and L. Emmons. 1984. Annotated checklist of bird and mammal species of Cocha Cashu Biological Station, Manu National Park, Peru. Fieldiana (Zoology) 21:1–29.

Terborgh, J., S. K. Robinson, T. A. Parker III, C. A. Munn, and N. Pierpont. 1990. Structure and organization of an Amazonian forest bird community. Ecological Monographs 60:213–238.

Thiollay, J.-M. 1992. Influence of selective logging on bird species diversity in a Guianan rain forest. Conservation Biology 6:47–63.

Thompson, F. R., III 1994. Temporal and spatial patterns of breeding Brown-headed Cowbirds in the midwestern United States. Auk 111: 979–990.

Thompson, F. R., III, T. M. Donovan, R. M. DeGraaf, J. Faaborg, and S. K. Robinson. 2002. A multi-scale perspective of the effects of forest fragmentation on birds in eastern forests. Studies in Avian Biology 25:8–19.

Vandermeer, J., I. G. de la Cerda, D. Boucher, I. Perfecto, and J. Ruiz. 2000. Hurricane disturbance and tropical tree diversity. Science 290: 788–791.

von Humboldt, and A. Bonpland. 1814–1829. Personal Narrative of Travels to the Equinoctial Regions of the New Continent, During the Years 1799–1804 [Title of first English edition, translated by H. M. Williams]. Longman, Hurst, Rees, Orme, and Brown, London.

Wallace, A. R. 1889. A Narrative of Travels on the Amazon and Rio Negro. Ward, Lock and Company, London.

White, G. C. 1996. NOREMARK: Population

estimation from mark–resighting surveys. Wildlife Society Bulletin 24:50–52.

White, G. C. 2002. Discussion comments on: The use of auxiliary variables in capture–recapture modelling. An overview. Journal of Applied Statistics 29:103–106.

White, G. C., and K. P. Burnham. 1999. Program MARK: Survival estimation from populations of marked animals. Bird Study 46 (Supplement): 120–138.

Wiens, J. A. 1977. On competition and variable environments. American Scientist 65:590–597.

Wilkinson, F. A., and U. R. Smith. 1997. The first nest records of the Sooty Antbird (*Myrmeciza fortis*) with notes on eggs and nestling development. Wilson Bulletin 109:319–324.

Williams, E. C., Jr. 1941. An ecological study of the floor fauna of the Panama rain forest. Bulletin of the Chicago Academy of Science 6:63–124.

Willis, E. O. 1966a. Interspecific competition and the foraging behavior of Plain-brown Woodcreepers, *Dendrocincla fuliginosa*. Ecology 47:667–672.

Willis, E. O. 1966b. The role of migrant birds at swarms of army ants. Living Bird 5:187–231.

Willis, E. O. 1967. The Behavior of Bicolored Antbirds. University of California Publications in Zoology, no. 79.

Willis, E. O. 1968. Studies of the behavior of Lunulated and Salvin's antbirds. Condor 70: 128–148.

Willis, E. O. 1969. On the behavior of five species of *Rhegmatorhina*, ant-following antbirds of the Amazon basin. Wilson Bulletin 81:363–395.

Willis, E. O. 1972. The behavior of Plain-brown Woodcreepers, *Dendrocincla fuliginosa*. Wilson Bulletin 84:377–420.

Willis, E. O. 1973. The behavior of Ocellated Antbirds. Smithsonian Contributions to Zoology, no. 144.

Willis, E. O. 1974. Populations and local extinctions of birds on Barro Colorado Island, Panama. Ecological Monographs 44:153–169.

Willis, E. O. 1976. Seasonal changes in the invertebrate litter fauna on Barro Colorado Island, Panama. Revista Brasileira de Biologia 36: 643–657.

Willis, E. O. 1978. Behavior and ecology of two forms of White-chinned Woodcreepers (*Dendrocincla merula*, Dendrocolaptidae) in Amazonia. Papeis Avulsos de Zoologia, São Paulo 33:27–66.

Willis, E. O. 1979. Ecologia e comportamento da mãe-de-taóca (*Phlegopsis nigromaculata*, Formicariidae). Revista Brasileira de Biologia 39:117–159.

Willis, E. O. 1982a. Ground-cuckoos (Aves, Cuculidae) as army ant followers. Revista Brasileira de Biologia 42:753–756.

Willis, E. O. 1982b. The behavior of Black-headed Antbirds (*Percnostola rufifrons*) (Formicariidae). Revista Brasileira de Biologia 42:233–247.

Willis, E. O. 1982c. The behavior of Scale-backed Antbirds. Wilson Bulletin 94:447–462.

Willis, E. O. 1984. The behavior of Black-banded Woodcreepers (*Dendrocolaptes picumnus*). Condor 84:272–285.

Willis, E. O. 1985. *Myrmeciza* and related antbirds (Aves, Formicariidae) as army ant followers. Revista Brasileira de Zoologia 2:433–442.

Willis, E. O., and Y. Oniki. 1978. Birds and army ants. Annual Review of Ecology and Systematics 9:243–263.

Willis, E. O., and Y. Oniki. 1992. As aves e as formigas de correição. Boletin Museo Paraense Emilio Goeldi 8:122–150.

Willson, S. K. 2000. First nest record of the White-throated Antbird (*Gymnopithys salvini*) and detailed nest records of the Hairy-crested Antbird (*Rhegmatorhina melanosticta*). Ornitologia Neotropical 11:353–357.

Wolda, H. 1978. Seasonal fluctuations in rainfall, food, and abundance of tropical insects. Journal of Animal Ecology 47:369–381.

Wright, S. J., C. Carrasco, O. Calderon, and S. Paton. 1999. The El Niño Southern Oscillation, variable fruit production, and famine in a tropical forest. Ecology 80:1632–1647.

Zimmer, K. J., and M. L. Isler. 2003. Family Thamnophilidae (Typical Antbirds). Pages 448-531 *in* Handbook of the Birds of the World, vol. 8. (J. del Hoyo, A. Elliott, and D. A. Christie, Eds.). Lynx Edicions, Barcelona, Spain.

Zyskowski, K., and R. O. Prum. 1999. Phylogenetic analysis of the nest architecture of Neotropical ovenbirds (Furnariidae). Auk 116:891–911.

APPENDIX

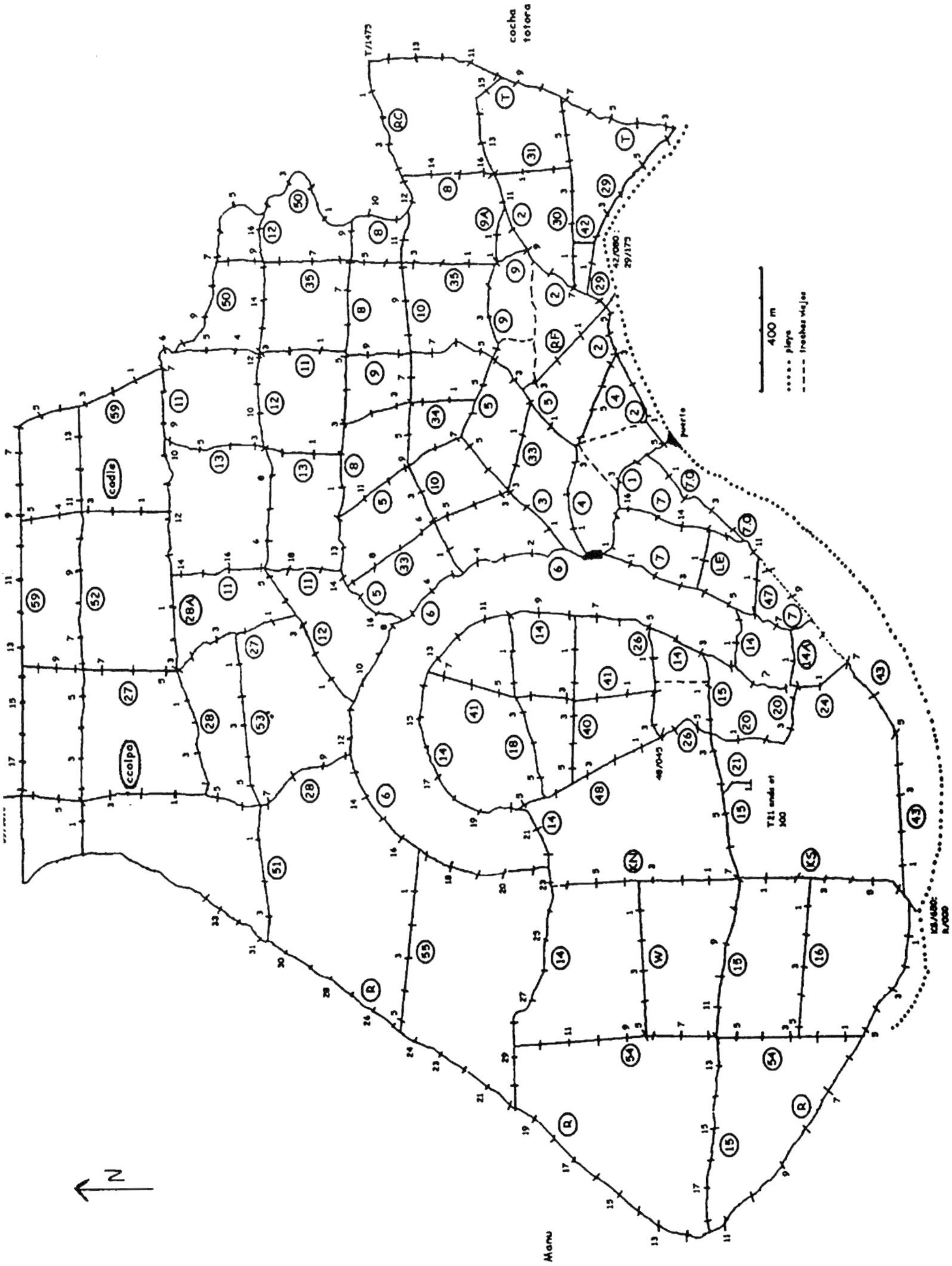

APPENDIX. Map of the trail system at Cocha Cashu Biological Station, Peru.